High Electrical Resistance Ceramics

Scrivener Publishing
100 Cummings Center, Suite 541J
Beverly, MA 01915-6106

Publishers at Scrivener
Martin Scrivener (martin@scrivenerpublishing.com)
Phillip Carmical (pcarmical@scrivenerpublishing.com)

High Electrical Resistance Ceramics

Thermal Power Plants Waste Resources

Muktikanta Panigrahi
Ratan Indu Ganguly
and
Radha Raman Dash

Scrivener Publishing

This edition first published 2023 by John Wiley & Sons, Inc., 111 River Street, Hoboken, NJ 07030, USA and Scrivener Publishing LLC, 100 Cummings Center, Suite 541J, Beverly, MA 01915, USA
© 2023 Scrivener Publishing LLC
For more information about Scrivener publications please visit www.scrivenerpublishing.com.

All rights reserved. No part of this publication may be reproduced, stored in a retrieval system, or transmitted, in any form or by any means, electronic, mechanical, photocopying, recording, or otherwise, except as permitted by law. Advice on how to obtain permission to reuse material from this title is available at http://www.wiley.com/go/permissions.

Wiley Global Headquarters
111 River Street, Hoboken, NJ 07030, USA

For details of our global editorial offices, customer services, and more information about Wiley products visit us at www.wiley.com.

Limit of Liability/Disclaimer of Warranty
While the publisher and authors have used their best efforts in preparing this work, they make no representations or warranties with respect to the accuracy or completeness of the contents of this work and specifically disclaim all warranties, including without limitation any implied warranties of merchantability or fitness for a particular purpose. No warranty may be created or extended by sales representatives, written sales materials, or promotional statements for this work. The fact that an organization, website, or product is referred to in this work as a citation and/or potential source of further information does not mean that the publisher and authors endorse the information or services the organization, website, or product may provide or recommendations it may make. This work is sold with the understanding that the publisher is not engaged in rendering professional services. The advice and strategies contained herein may not be suitable for your situation. You should consult with a specialist where appropriate. Neither the publisher nor authors shall be liable for any loss of profit or any other commercial damages, including but not limited to special, incidental, consequential, or other damages. Further, readers should be aware that websites listed in this work may have changed or disappeared between when this work was written and when it is read.

Library of Congress Cataloging-in-Publication Data

ISBN 978-1-394-19993-8

Cover image: Pixabay.Com
Cover design by Russell Richardson

Set in size of 11pt and Minion Pro by Manila Typesetting Company, Makati, Philippines

Printed in the USA

10 9 8 7 6 5 4 3 2 1

Contents

Preface   ix

1 Fundamentals of Thermal Power Plant Wastes-as Ceramic Backbone   1
     1.1 Introduction   1
     1.2 Thermal Power Plant Wastes   2
     1.3 Generation of Thermal Power Plant Ashes   2
     1.4 Nature and Composition of Thermal Power Plant Ashes   5
     1.5 Characteristics of Thermal Power Plant Ashes   10
     1.6 Causes of Resistance in Insulator   14
     1.7 Resistance Measurement   15
     1.8 Different Methods for Resistivity Measurement   15
     1.9 Resistance Temperature Detector (RTDs)   19
         1.9.1 Benefit of RTD   19
     1.10 Platinum Resistance Thermometer (PRTs)   19
     1.11 Thermal Power Plant Wastes (i.e., Coal Ash) Management   20
     1.12 Literatures Survey on Thermal Power Plant Wastes-Based Ceramics   20
     1.13 Conclusions   24
     Acknowledgements   25
     References   25

2 Ceramic Production Methods and Basic Characterization Techniques   35
     2.1 Introduction   35
     2.2 Characterization Techniques   41
         2.2.1 X-Ray Diffraction (XRD) Technique   41
         2.2.2 Fourier Transformation Infra-Red (FTIR) Spectroscopy   42
         2.2.3 Scanning Electron Microscopy (SEM)   42
         2.2.4 Electrical Characterizations   43

		2.2.4.1	Electrical Resistivity Measurement by Two Probe (at Room Temperature)	43

	2.3 Conclusions	44
	Acknowledgements	45
	References	45
3	**High Resistance Sintered Fly Ash (FA) Ceramics**	**51**
	3.1 Introduction	52
	3.2 Experimental Details	53
	3.2.1 Materials and Chemicals	53
	3.2.2 Materials Preparation	54
	3.2.3 Physical Characterizations	55
	3.2.4 Results and Discussion	56
	3.3 Conclusions	61
	Acknowledgements	61
	References	61
4	**High Resistance Sintered Fly Ash/Kaolin (FA/CC) Ceramics**	**65**
	4.1 Introduction	66
	4.2 Experimental Section	66
	4.2.1 Materials and Method	66
	4.3 Preparation of Test Samples	67
	4.3.1 Preparation of Sintered FA/CC Composite	67
	4.4 Characterization Techniques	68
	4.5 Results and Discussion	69
	4.6 Conclusions	75
	Acknowledgements	75
	References	75
5	**High Resistance Pond Ash Geopolymer Ceramics**	**79**
	5.1 Introduction	80
	5.2 Experimental Details	83
	5.2.1 Materials and Chemicals	83
	5.3 Test Methods	85
	5.4 Results and Discussion	86
	5.5 Conclusions	92
	Acknowledgements	92
	References	92
6	**High Resistance Sintered Pond Ash Ceramics**	**97**
	6.1 Introduction	98
	6.2 Experimental Details	99
	6.2.1 Materials and Chemicals	99

		6.2.2	Materials Preparation	99

		6.2.2	Materials Preparation	99
		6.2.3	Test Methods	101
		6.2.4	Results and Discussion	102
	6.3	Conclusions		108
		Acknowledgements		109
		References		109
7	**High Resistance Sintered Pond Ash/Kaolin (PA/CC) Ceramics**			**115**
	7.1	Introduction		115
	7.2	Experimental Details		118
		7.2.1	Materials and Chemicals	118
		7.2.2	PA/Kaolin Composite Preparation	118
		7.2.3	Test Methods	121
	7.3	Results and Discussion		122
	7.4	Conclusions		130
		Acknowledgements		131
		References		131
8	**High Resistance Sintered Pond Ash/Pyrophyllite (PA/PY) Ceramics**			**137**
	8.1	Introduction		137
	8.2	Experimental Section		140
		8.2.1	Materials and Chemicals	140
	8.3	Preparation of PA/PY Composite Materials		140
	8.4	Test Methods		143
	8.5	Results and Discussion		144
	8.6	Conclusions		152
		Acknowledgements		152
		References		153
9	**High Resistance Sintered Pond Ash/k-Feldspar (PA/k-FD) Ceramics**			**159**
	9.1	Introduction		160
	9.2	Experimental Details		162
		9.2.1	Materials and Chemicals	162
	9.3	Preparation of PA/FD Sintered Materials		162
	9.4	Test Methods		165
	9.5	Results and Discussion		166
	9.6	Conclusions		174
		Acknowledgements		175
		References		175

10 Applications, Challenges and Opportunities of Industrial Waste Resources Ceramics — **181**

- 10.1 Introduction — 182
- 10.2 Different Ways of Utilization of Waste — 183
  - 10.2.1 Porous Insulation Refractory — 183
  - 10.2.2 Dense Refractory — 183
  - 10.2.3 Ceramic Tiles — 184
- 10.3 Glass — 187
- 10.4 Glass-Ceramic (GC) — 188
- 10.5 Mullite — 188
- 10.6 Wollastonite — 188
- 10.7 Cordierite — 189
- 10.8 Silicon Carbide — 189
- 10.9 Silicon Nitride — 190
- 10.10 Ceramic Membranes — 190
- 10.11 Challenges — 190
- 10.12 Opportunity — 191
- 10.13 Conclusions — 191
- Acknowledgments — 192
- References — 193

Index — 199

Preface

The pond and fly ash generated by thermal power stations are waste materials which pollute the environment. Dumping these materials on land spoils its cultivability. Therefore, the present challenge for scientists and researchers is the fruitful utilization of these materials. This will serve the dual purpose of minimizing environmental pollution and conserving land for cultivation.

The major constituents of fly ash (FA), pond ash (PA), and bottom ash (BA) are SiO_2, Al_2O_3, and Fe_2O_3. Their composition, however, will depend on the nature of coal used by thermal power plants. The present problems associated with these materials can be solved by minimizing power generation through thermal routes with alternative power generation methods such as solar energy, nuclear energy, wind energy, hydraulic electricity, etc. However, in reality, due to increasing urbanization and industrialization, power requirements are also increasing, which are being met by thermal power generation plants. Hence, the production of these materials (FA/PA/BA) is inevitable. Therefore, investigations are currently being conducted in an effort to utilize pond and fly ash to produce a novel material called ceramics, in which kaolin/pyrophyllite/feldspar are mixed with pond/fly ash. These thermal power plant waste-based ceramics can replace porcelain-based ceramics. Since permanent ceramic components can be developed using these wastes, it is no wonder why some scientists are trying to develop ceramics utilizing them.

This book highlights the preparation of ceramics using pond/fly ash. Since the mullite phase formed by heat treatment improves the properties of ceramics, current investigations will perhaps be the first attempt to develop ceramics using pond ash. The properties of components made with these developed ceramics are found to be comparable to those made with porcelain. Since evidence has shown the formation of mullite after heat treatment, systematic investigations are being carried out to understand phase transformation during thermal treatment. Upon the performance of these materials above ambient temperature being evaluated, results have indicated the possible replacement of porcelain with these newly invented ceramics prepared with pond ash.

The extensively reviewed chapters of this book illustrate the current status of research on these materials. At the end of each of the 10 chapters, conclusions are drawn which will benefit researchers working in this area. Chapter 1 discusses the fundamentals of thermal power plant wastes, which are the backbone of ceramics. Different production methods of ceramics and various characterization techniques are discussed in Chapter 2. This will help new researchers to progress in further development of new ceramics from waste on a laboratory scale. Chapter 3 describes the preparation of ceramics from fly ash. The preparation of ceramics from fly ash/kaolin composite is discussed in Chapter 4; and Chapter 5 is devoted to the production of ceramics using pond ash. Chapter 6 describes the preparation and characterization of geopolymer from pond ash; and the preparation of pond ash composite (pond ash and kaolin) is described in Chapter 7. Chapter 8 deals with the production of ceramic matrix composite (CMC) using pond ash and pyrophyllite, while Chapter 9 discusses the preparation of ceramics using pond ash and k-feldspar mixture. The book concludes with a discussion of the applications, challenges and opportunities regarding ceramics from industrial waste resources in Chapter 10.

The Authors
June 2023

1

Fundamentals of Thermal Power Plant Wastes-as Ceramic Backbone

Muktikanta Panigrahi[1]*, Ratan Indu Ganguly[2] and Radha Raman Dash[3]

[1]*Department of Materials Science, Maharaja Sriram Chandra BhanjaDeo University, Balasore, Odisha, India*
[2]*Department of Metallurgical Engineering, National Institute of Technology, Raurkela, Odisha, India*
[3]*CSIR-National Metallurgical Laboratory, Jamshedpur, Jharkhand, India*

Abstract

The chapter has provided extensive reviews on production of fly ash/pond ash/bottom ash globally. Properties of different minerals present in fly ash are discussed. By adjustment of chemical composition of fly ash, useful dielectric materials can be developed by proper treatment. Since compositions of pond ash, bottom ash, and fly ash depend on resource material like coal, therefore, judicial selection of compositions and their adjustment is very important for developing dielectric materials.

Keywords: Energy, thermal power point waste, fly ash, bottom ash, pond ash, ceramics, electrical resistance measurement, two point method, four point method, Van-der Pauw method, thermal waste managements

1.1 Introduction

Energy is the backbone of a country. Sustainable development and growth of industries largely depend on energy. Therefore, it plays a vital role for the economic development of a nation. A developing country like India is still dependent on natural reserves like coal. Even in the 21st century, coal is the first choice as a fuel for electricity generation [1]. In any coal based thermal power plants, Coal fly ash generated is considered to be a

---

*Corresponding author: muktikanta2@gmail.com

Muktikanta Panigrahi, Ratan Indu Ganguly and Radha Raman Dash. *High Electrical Resistance Ceramics: Thermal Power Plants Waste Resources*, (1–34) © 2023 Scrivener Publishing LLC

by-product [2]. Presence of fly ash in air (in micro fine level) affects human health. However, residual burnt product i.e. ash is considered to be a waste. Dumping of ash into a pond saves air pollution but at the same time it causes water pollution. Additionally, some of the undesirable elements such as Sb, As, Pb etc cause problems for living creatures in water.

Due to advancement in technologies, environmental pollution is enhanced. In addition, there is global warming which affects the climate and environment. Therefore, the Government is making policies to control pollution by consuming wastes for development of useful products.

Financial support need be sanctioned to develop value-added products from these wastes i.e., fly ash, bottom ash and pond ash.

Based on their mineralogical properties, these materials can be used for developing tiles, zeolite, geopolymer, etc.

In the present chapter discussion has been made on coal-based thermal power plants, ash production, properties, and utilization.

1.2 Thermal Power Plant Wastes

Due to rapid urbanization and industrialization in developing countries [3], natural reserves are being exhausted to be increasing demand for production of energy. In 21st century, coal is the prime choice for generating electricity. During electricity production in thermal power plants, Coal ash is produced as waste [2]. Waste increases environmental pollution. It is to ensure that sustainable growth and development of urban cities need a receptive, innovative, and productive environment for a better quality of life.

1.3 Generation of Thermal Power Plant Ashes

Fly ash is generated from thermal power plant and steam producing plants [4]. Normally, coal is blown with air in ignition chamber. Quickly, it is heated where there is formation of liquid which causes deposit of mineral. However, the pipe gas and creating the liquid mineral builds up to solidify and structure slag. Coarse particles present in coal stay back in bottom of the heating chamber and suspended particles are blown with air to the atmosphere. This hot blown air contains fine particles of minerals with unburnt carbon are passed through an electrostatic precipitator where particles get settles. These are called fly ash [4]. Liquid minerals form which solidifies resulting slag.

Optical image and SEM image of Fly Ash is shown in Figure 1.1.

Ash stored in a heating chamber is a non-combustible residue. It is likely that some traces of combustibles are embedded in clinkers. Clinkers are to be stick to hot side walls of a coal-burning furnace. Clinkers fall by themselves into bottom hopper of a coal-burning furnace and are cooled. The above portion of ash is known as bottom ash [7]. Optical image and SEM image of Bottom Ash is shown in Figure 1.2.

Pond ash [10] is a waste collection of ash from burning chamber of boilers. In short, it can be defined as wet disposal of fly ash mixed with bottom ash. Pond ash is obtained from in large ponds or dykes as slurry. Pond ash is increasing in an alarming rate. Its production ash is posing threat to our environment. Therefore, sustainable management of these materials has become a thrust area in engineering research. Optical image and SEM image of Pond Ash are shown in Figure 1.3.

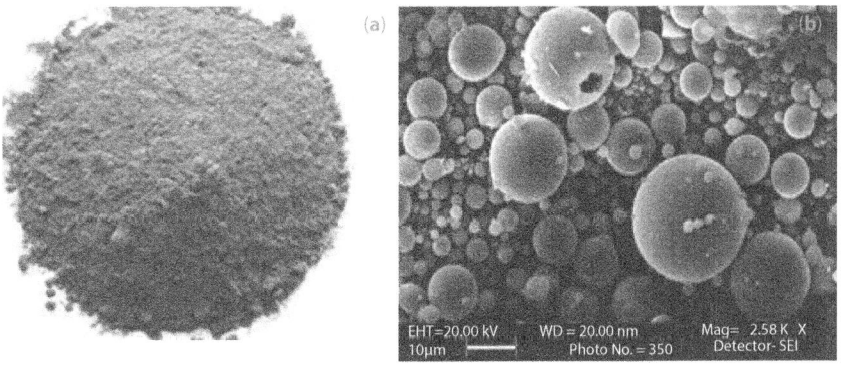

Figure 1.1 Optical micrograph of Fly ash (a) [5] and SEM image of Fly Ash (b) [6].

Figure 1.2 Optical image of Bottom Ash (a) [8] and SEM image of Bottom Ash (b) [9].

4 HIGH ELECTRICAL RESISTANCE CERAMICS

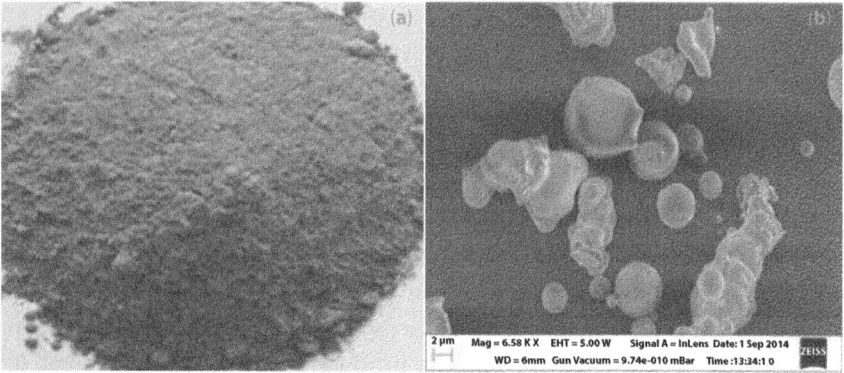

Figure 1.3 Optical image of Pond Ash (a) [11] and SEM image of Pond Ash (b) [12].

Grading of coal is made on the basis of ash content [13]. Usually, coal used in thermal power stations, contains ash in the ranges 30-40% (with the average ash content around 35%). However, for metallurgical industries, different grade coal is used. Metallurgical coal should contain low amount of ash and high clerking index. Generation of ash at the power plants is shown schematically in Figure 1.4.

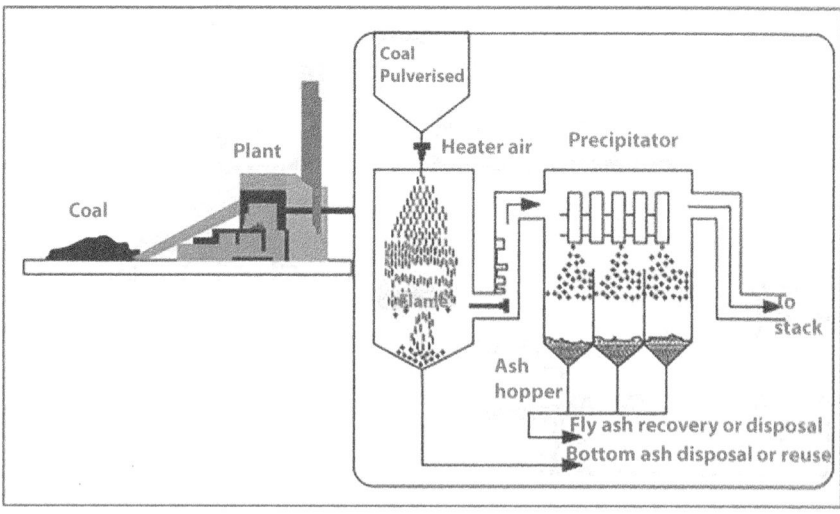

Figure 1.4 Generation of ash at the power plants [14].

1.4 Nature and Composition of Thermal Power Plant Ashes

Micron-sized Fly ash particle is collected as flue gas from gases of coal fired plants during the production of electricity by electrostatic precipitators. Table 1.1 describes composition of coal (Bituminous, Subbituminous, and Lignite). It is revealed from Table 1.1 that major constituents of coal are silica, alumina, iron oxide, and lime. However, other constituents (magnesia, sulphur oxide, sodium oxide, and potassium oxide) are comparatively less than 10%.

In Fly ash [14], Elements with major percentages are silica, alumina and iron, whereas minor percentage elements are sodium, potassium, titanium, etc. Also, large amounts of non-combustible impurities (i.e., limestone, shale, dolomite, feldspar and quartz) are main constituents of coal ash obtained from thermal power plant. Chemical composition of fly ash using different types of coal is shown in Table 1.1.

ASTM C618 classifies fly ash into two classes i.e., Class F fly ash and Class C fly ash (Table 1.2). This classification is based on amounts of content i.e., calcium, silica, alumina, and iron (Table 1.2). Chemical properties of fly ash are largely depends on its chemical compositions and occurrence of this minerals [15].

Table 1.1 Normal range of chemical compositions for Fly ash produced from different coal types [14].

S. no.	Component	Bituminous	Subbituminous	Lignite
1	SiO_2 (%)	20–60	40–60	15–45
2	Al_2O_3 (%)	5–35	20–30	20–25
3	Fe_2O_3 (%)	10–40	4–10	4–15
4	CaO (%)	1–12	5–30	15–40
5	MgO (%)	0-5	1-6	3-10
6	SO_3 (%)	0-4	0-2	0-10
7	Na_2O (%)	0-4	0-2	0-6
8	K_2O (%)	0-3	0-4	0-4
9	LOI	0-15	0-3	0-5

Table 1.2 Typical composition of Class F Fly Ash, Class C Fly Ash, and Portland cement with their ASTM standard.

Components	Class F Fly Ash		Class C Fly Ash		Portland cements	
	Typical*	ASTM C-618	Typical**	ASTM C-618	Typical***	ASTM C-150
SiO_2	48.0	---	37.3	---	20.25	---
Al_2O_3	24.3	---	21.4	---	4.25	---
Fe_2O_3	15.6	---	5.7	---	2.59	---
$SiO_2 + Al_2O_3 + Fe_2O_3$	87.9	70.0 (min.%)	64.3	50.0 (min. %)	---	---
CaO (Lime)	3.2	---	22.4	---	60.6	---
MgO	---	---	---	---	2.24	6.0 (max.%)
SO_3	0.4	5.0(max.%)	2.5	5.0(max.%)	---	3.0 (max.%)
Loss of Ignition (LOI)	3.0	6.2 (max.%)	0.4	6.0 (max.%)	0.55	3.0 (max.%)
Moisture content	0.1	3.0	0.1	3.0	---	---
Insoluble residue	---	---	---	---	---	0.75
Available alkalis as equivalent Na_2O	0.8	1.5	1.4	1.5	0.20	---

Class F fly ash

Class F fly ash is generated from coal containing less than 7% lime (CaO). It is pozzolanic in nature [16]. Because of pozzolanic behaviour, Class F fly ash is used as a cementing agent, such as Portland cement. Class F ash contains alkaline activator such as sodium silicate and alkali and therefore, it can form a geopolymer. Geopolymer is a cementitious material.

Class C fly ash

Class F fly ash is obtained from burning of lignite or sub-bituminous coal in thermal power stations. Class F fly ash has self-pozzolanic behaviour. Since, it contains higher amount of limes [16]. Therefore, it does not require an activator for producing cementitious materials. Also, alkali and sulfate contents are generally higher in Class C fly ashes. Composition of Class F Fly Ash, Class C Fly Ash, and Portland cement is presented in Table 1.2.

Coal bottom ash is coarse and granular, nature. Bottom ash is collected from bottom of furnaces. Usually, a coal-burning furnace is a dry in nature. Out of 100% ash, 20 percent of ash is dry bottom ash. They look is a dark grey, granular, porous, sand size 12.7 mm (½ in). They usually is collected in a water-filled hopper locked at the bottom of the furnace [17].

Main constituents of bottom ash are silica, alumina, and iron, whereas calcium, magnesium, sulphates. However, alkali constituents such as Ca, Mg, Na, and K are present in a trace amount. Also, it contains very less amount of sulphur. Table 1.3 presents a chemical analysis of bottom ash. If bottom ash is produced from burning of lignite or sub-bituminous coals, it has higher percentage of calcium if compared with bottom ash produced from burning of anthracite or bituminous coals. Bottom Ash show corrosive properties because of salt content and low pH value. Corrosively indicator tests normally used to evaluate bottom ash are pH, electrical resistivity, soluble chloride content, and soluble sulfate content. Bottom ash is found to be non-corrosive if pH exceeds 5.5 and electrical resistivity is greater than 1,500 ohm-centimetres. Other criterion for defining non-corrosive nature is soluble chloride content. If bottom ash is less than 200 parts per million (ppm) soluble chloride content or less than 1,000 parts per million (ppm) soluble sulfate content, then it is non – corrosive [18].

Table 1.3 Chemical composition of Bottom Ash (BA) [19].

Raw materials	SiO_2	Al_2O_3	CaO	MgO	Fe_2O_3	TiO_2	Cr_2O_3	MnO	P_2O_5	C	LOI
Bottom Ash	68.0	25.0	1.66	0.02	2.18	1.45	0.00	0.00	0.00	0.0	1.69

Microstructural aspect of Bottom ash show angular particles with a very porous surface texture. Bottom ash particle size is ranging between sizes of fine gravel to fine sand with very low percentages of silt-clay sized particles.

Specific gravity of dry bottom ash is depends on its chemical composition. If unburnt carbon is more, then specific gravity value is lower. Bottom ash with a low specific gravity has a porous or vesicular texture. They easily degrade under loading or compaction [20]. Table 1.4 indicates some typical physical properties of bottom ash.

Pond ash [22] is obtained from wet disposal of fly ash mixed with bottom ash. Since it is obtained from large ponds or dykes as slurry. It is called pond ash. Pond ash is being generated at an alarming rate. Therefore, its sustainable management has become the thrust area in engineering research. Pond ash is relatively coarse and dissolvable in alkalis. Since it is obtained from water, and hence it's pozzolanic reactivity becomes low. Due to less pozzolanic effect, it is not preferred as replacement of cement in concrete. Optical image and SEM image of Pond Ash is shown in Figure 1.5.

There is a possibility to use pond ash for making burnt clay bricks. In bricks production at brick manufacturing plants, pond ash is mixed with clay to manufacture bricks using conventional methods. Usually, green bricks are fired (traditional way) to produce brick products. Different tests (i.e., tolerance in dimension, water absorption, compressive strength, initial rate of absorption and weathering) are performed to evaluate performance of bricks. Ash ponds use gravity to settle out large particulates from power plant waste water. This technology does not treat dissolved pollutants [23]. Ponds generally have not been built as lined landfills, and therefore chemicals in the ash can leach into groundwater and surface waters, accumulating in the biomass of the system [24–26]. Chemical composition of pond ash is presented in Table 1.5. Physical characteristics of pond ash are indicated in Table 1.6.

Table 1.4 Some typical physical properties of bottom ash [21].

S. no.	Different type physical property	Physical property value	Reference
1	Specific gravity	2.1 - 2.7	6 (bottom ash)
2	Dry unit weight	720 - 1600 kg/m3	6 (bottom ash)
3	Plasticity	(45 - 100 lb/ft3)	6 (bottom ash)
4	Absorption	none	4 (bottom ash)

Table 1.5 Chemical composition of Pond Ash (PA) [11].

Raw materials	SiO_2	Al_2O_3	CaO	MgO	Fe_2O_3	TiO_2	Cr_2O_3	MnO	P_2O_5	C	LOI
Pond Ash	62.8	28.3	0.7	0.58	3.85	1.84	0.04	0.03	0.32	1.15	0.5

Table 1.6 shows chemical and physical characteristics of pond ash obtained from other regions. There is difference in composition of pond ash (Table 1.5). Alumina and silica content (Table 1.5) is 91%, whereas chemical composition of pond ash (Table 1.5) is 79 %. The other major constituent (Table 1.5) is Iron oxide (Fe_2O_3) and TiO_2. Both the compounds are higher percentage in Table 1.5. Other important feature of Table 1.5 is higher percentage of alkali content.

It is concluded that type of coal, performance of generating facility, variety collection, disposal & storage methods, coal burning temperature, and peak load demand in thermal stations, etc are controlled properties of pond ash. Physical, chemical, and mineralogical characteristics of pond ash play an important role in different sectors. Therefore, use of pond ash has to done judicially depending on the nature of applications.

Table 1.6 Chemical and physical characteristics of pond ash.

Chemical characteristics [11]		Physical characteristics [27]	
Parameters	Concentration (% by wt.)	Parameter	Pond ash
SiO_2	59.007		
Al_2O_3	19.551	Specific gravity @ 27°C	2.0675
Fe_2O_3	15.350	Fineness (m2/kg)	73.78
TiO_2	3.158	Hydraulic conductivity @ 27°C	0.992
K_2O	1.271	Dry density, γd (g/cc)	0.848
CaO	1.151	Void ratio	1.435
Mn_2O_3	0.197		

10   HIGH ELECTRICAL RESISTANCE CERAMICS

1.5   Characteristics of Thermal Power Plant Ashes

Specific gravity is an important physical property that is needed to use coal ashes in different applications, particularly geotechnical areas [28]. Generally, specific gravity of coal ashes varies from 1.6 to 3.1. Variation of specific gravity causes many factors such as gradation, particle shape and chemical composition. Low specific gravity coal ashes may be due to the presence of a large number of hollow cenospheres. Hollow cenospheres entraps air, and also varies in chemical composition (particularly iron content) in coal from which it is obtained.

Fly ash has usually higher specific gravity compared to pond and bottom ashes (obtained from the same locality). When the particles are crushed, they show a higher specific gravity compared to the uncrushed portion of the same material [29].

Specific surfaces [28] of coal ashes are important characteristics and supports to understand their physical and engineering behaviour. Coal ashes are primarily silt/sand-sized particles and their specific surface is expected to be very low.

Surface area of particles is important because it may control total adsorption capacity but not necessarily the desorption rate. Surface areas of fly ash particles generally vary inversely with the particle size. Smaller the particle size, larger the surface area [30].

Permeability or Hydraulic Conductivity (K) is often used interchangeably for same property. Coefficient of permeability or hydraulic conductivity (K) takes into account of fluid properties, whereas intrinsic permeability (k) only refers to effectiveness of porous medium alone [31]. They are related by following equation 1.1;

$$K = k \times \text{specific weight of fluid/dynamic viscosity of fluid} \quad (1.1)$$

Fluid is actually flowing through the void spaces, not particulate matter. Therefore, porosity can have a controlling influence on permeability. Porosity is a value that portrays the amount of voids in a sample which is representative of water bearing capacity. Porosity is usually represented by 'n'. It is determined by calculating the ratio between volumes of voids and total volume multiply by 100 (to express as percent). Obviously, porosity and permeability can be affected by compaction (density) since this reduces amount of void space for a given total volume. Normally, higher porosity samples will have a higher hydraulic conductivity. Fly ash compacted in a laboratory to 95 % maximum density can achieve a permeability of

1×10^{-5} cm/sec. A higher density results in a lower permeability. This is beneficial since a low permeability will restrict leachate from migrating away from the site.

When fly ash is used to change soil for plant growth, the hydraulic conductivity in the soil increases upto 10 to 20 percent by volume of fly ash. Beyond this point, hydraulic conductivity of soil decreases. This seems to be a pozzolanic reaction occurring in fly ash which tends to cement in contact with water.

Density is also affected by compaction. In a bituminous coal fly ash, a 95 percent maximum density (1.3 g/cm³) is achieved [28]. Since fly ash generally has a low bulk density, fly ash addition to soil reduces the bulk density of soil.

Grain size distribution [32] is indicated by well graded/poorly graded, fin/course, etc. It helps in classifying coal ashes. Coal ashes are predominantly silt sized with some sand-size fraction. Leonard's and Bailey [33] have reported the range of gradation for fly ash and bottom ashes which can be classified as silty sands or sandy silts.

In Indian coal ashes, fly ashes consist predominantly of silt-size fraction mixed with some clay-size fraction. Pond ashes consist of a silt-size fraction with some sand-size fraction. Bottom ashes are coarser particles consisting predominantly of sand-size fraction mixed with some silt-size fraction. Based on the grain-size distribution, coal ashes can be classified as sandy silt to silty sand. They are poorly graded with coefficient of curvature ranging between 0.61 and 3.70. Coefficient of uniformity is in the range of 1.59-14.0.

Free swell index [34] is a tool to identify swelling behaviour of soils. Free swell test method has been proposed by Holtz and Gibbs to estimate swell potential. Sridhar *et al.* [35] have modified the definition of free swell index itself to take care of the limitations. 70% coal ashes show negative free swell index which is due to flocculation. Since clay-size fraction in coal ashes is very less, and hence, free swell index is negligible. Index properties [28] are extensively used in geotechnical engineering practice.

Percussion cup and fall cone methods are popular to determine the liquid limit of fine-grained soils. In Percussion cup method, it is very difficult to cut a groove in soils of low plasticity. It has a tendency to slip rather than flow. Fly ashes are not studied because of their non-plastic nature. In cone penetration method, fly ash is not been studied because of variation of water content in the cup with depth. So, it is very difficult to get a smooth surface of fly ash in the cup.

A new method, Equilibrium water content under K_o stress method has been found to determine liquid limit of coal ashes except class C fly ashes.

This method is simple, reasonably error free, less time consuming and good reproducibility.

Chemical properties of coal ashes greatly influence an environmental impact. It may arise due to their use/disposal/engineering properties. The adverse environmental impacts include contamination of surface and sub-surface water with toxic heavy metals, loss of soil fertility around the plant sites, etc. which is present in coal ashes. Hence detailed studies of these ashes include their chemical compositions, morphologies studies, pH, presence of total soluble solids, etc are essential [28].

Chemical compositions determine the possible applications of coal ash in different sectors. Particularly, Indian coal ashes satisfy chemical requirements for their use (as a pozzolanic). According to ASTM classification, only Neyveli fly ash can be classified as Class C fly ash and all other coal ashes fall under Class F [28].

pH [36] of aqueous medium affects physico-chemical properties. Further, mobilization of trace elements in aqueous medium regulates solubility of hydroxide and carbonate salts which depending upon pH of aqueous media. Fly ash has higher pH values in comparison to pond and bottom ashes. This is because of the presence of higher amounts of free lime and alkaline oxides. Since coal ashes are nearly alkaline. Therefore, it can be used in reinforced cement concrete. This will be safe against corrosion.

Presence of soluble solids is an important aspect and has greatly influenced engineering properties. Further, solubility of nutrient elements like calcium, magnesium, iron, sulphur, phosphorus, potassium and manganese enhance crop yield to a great extent. Soluble solids range is found to be 400-17600 ppm (for fly ashes), 800 - 3600 ppm (for pond ashes), and 1400 - 4100 ppm (for bottom ashes), respectively.

Strength of fly ash generally improves with time due to pozzolanic reactions [37]. Reactive silica and free lime contents are necessary for pozzolanic reactions. Lime reactivity is a property which depends on the proportion of reactive silica in coal ash. It is also found to be high for fly ash as compared to bottom and pond ashes. Higher percentage of free lime in coal ash plays an important role in lime reactivity.

In Compaction behaviour [38], density of coal ashes is important and controls strength, compressibility and permeability properties. Densification of ash improves engineering properties. Compactions of material depend on the amount of ash used, method of energy application, grain size distribution, plasticity characteristics and moisture content. Variation of dry density with moisture content of fly ash is less than that of a well-graded soil (if both have the same median grain size). Fly ash is to be less sensitive to the variation of moisture content in comparison to soils.

It could be explained by the presence of void in fly ash. Normally, soils have air voids and dry density is ranging from 1 - 5%, whereas dry density of fly ash is ranging from 5 - 15%. The higher void content could tend to limit the build-up of pore pressures during compaction, thus fly ash is to be compacted over a larger range of water content.

Strength behaviour [39] is an important engineering property. It is essential for many geotechnical applications. Compressive strength properties of fine ashes are higher than those of coarser ashes based specimens. Also, most ash specimens have higher shear strength due to internal friction. Effects of additives such as gypsum and lime on ashes based specimens have shown increased strength. In some cases, addition of gypsum on ashes based specimens has no effects on the strength characteristics.

Permeability [40] is an important parameter in design liners. Coefficient of permeability of ash depends upon grain size, degree of compaction and pozzolanic activity. Permeability is found to be 8×10^{-6} cm/s - 1.87×10^{-4} cm/s for fly ash, 5×10^{-5} cm/s - 9.62×10^{-4} cm/s for pond ash, and 9.9×10^{-5} cm/s - 7×10^{-4} cm/s for bottom ash.

Leaching behaviour
Leaching behaviour can be defined as passing of water soluble components of matrix through a porous media [41]. Permeation of contaminated pore is a driving force, called leaching. Contamination of water is due to solution of the soluble constituents of matrix and consequent water passes through a porous matrix called leachate. Capacity of removal of leach – out material through matrix is called leachability. Particularly, fly ash may contain various toxic elements. Leaching of these toxic elements from ash to pond water is gaining considerable importance. It may cause serious environmental problems. Release of contaminants from fly ash and their subsequent influence on ground-water quality is governed by several factors including quality of coal, sources of water, pH, time, temperature, etc. Leachate characteristics are highly variable and even within a given landfill site, leachate quality varies over time and space [41].

Electrical Properties
In general, the most common properties of ceramics are: hardness, wear-resistance, brittleness, refractoriness, thermal insulating property, electrical properties, nonmagnetic, oxidation resistance, thermal shock prone and chemically inertness [42]. Nowadays, ceramics are used in various sectors. Properties of ceramics mainly depend on types of atoms present, nature of bonding (between the atoms) and type of packing. In ceramic materials, atoms are wide-spread through a chemical bond. In the materials, two types

Figure 1.5 Schematic diagram of a simple resister [47].

of chemical bonds (i.e., covalent and ionic) are indicated. Since free electrons are bound with neighboring atoms, therefore, no free electrons are available to carry electricity. Hence, ceramics are insulators unlike metal [43]. Hence, ceramic materials are dielectrics and support electrostatic fields. In ceramics, electrical conductivity varies with frequency (of the applied field) and temperature [44]. This will provide information about the charge transport phenomenon. Ceramic materials are employed for manufacturing capacitors, insulators and resistors especially for electronic devices [45].

Modern concept of electrical resistance was first discovered by G. S. Ohm (1926) [46]. Formulated equation between voltage and current and the expression 1.2 is as [46];

$$R = \left(\frac{V}{I}\right) \quad \ldots\ldots\ldots\ldots\ldots\ldots(1.2) \text{ (Ohms law)}$$

Here; R=resistance of material in ohm
  V=Obtained voltage in volt
  I=Applied current in Ampere

Unit (SI) of electrical resistance is ohm (Ω). The reciprocal of resistance is termed as electrical conductance and its unit is Siemens. Figure 1.5 shows schematic diagram of a simple resister.

1.6 Causes of Resistance in Insulator

In insulator [48], electric current does not flow freely. This happens due to tightly bound electrons in an atom and, hence electrons are unable to move freely. Resistivity is a key property which is used to distinguish electrical materials. Therefore, Insulators have higher resistivity if compared with other electrical materials.

Electrical insulation is due to absence of electrical conduction. According to electronic band theory [49], numbers of available states are observed in an insulator. Insulators have large band gaps. Such type of insulator is known as Mott insulator. In Mott insulator, there are electrons in the valence band and none of the electrons are in the conduction band. There is a large energy gap between the bands i.e., greater than 3 eV (electron volt). So, conduction is not possible.

1.7 Resistance Measurement

Quantitative measurement of a material's opposition to the flow of current is named resistivity. It depends only on the composition of material (and not on the shape and size). Figure 1.6 Schematic diagram for resistivity measurement Resistance is calculated by the following equation 1.3 [50];

$$R = \frac{\rho l}{A} \quad \text{...........(1.3)}$$

Where,
R is the resistance (ohms)
ρ is resistivity (ohm-meters)
L is the length (meters)
A is the cross-sectional area (square-meters)

1.8 Different Methods for Resistivity Measurement

Resistivity is determined by measuring resistance R and the dimensions of the sample (length l, width/thickness d). Resistance (R) is generally measured by a voltage-current method. A known current "I" is applied into the sample and voltage "V" is determined by point contact. From this

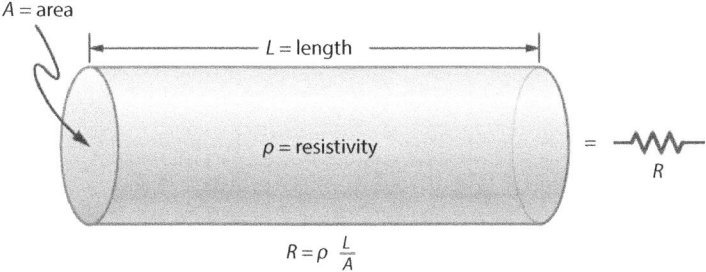

Figure 1.6 Schematic diagram for resistivity measurement [50].

measurement, resistivity (ρ) is estimated using the following relation ie, equally [50];

$$\rho = \left(\frac{Rw \times d}{l}\right) \quad \ldots\ldots\ldots\ldots(1.4)$$

Different methods are adapted to calculating resistivity. They are discussed one by one.

Two probe method
Schematic diagram of two probes electrical resistivity measurement [51] is shown in Figure 1.7. In two probe methods (Figure 1.7), test specimens may be strip-, rod- or bar-type. In this method (Figure 1.7), probes always contribute from contact wires. Contact resistances are therefore, chosen from among method which have low resistance like metals.

Since, the internal resistance of the voltmeter is very high (106 Ω). Therefore, voltage measured by the voltmeter will be calculated by the following expression 1.5 (I0<<I, so that I-I_0 ~ I) [52];

$$V = (I - I_0)(r + R + r) \quad \ldots\ldots\ldots\ldots(1.5)$$

So, r+R+r give an error to the measurement. This method is adopted for highly resistive materials where R>>2r.

Four probe method
Schematic diagram of four probes electrical resistivity measurement is shown in Figure 1.8. Four probe method support to minimize contributions of lead resistance, contact resistance, etc. (Figure 1.8). This method is accurate for measuring resistance of sample. In this method, four probes are contacted linearly with equally spaced. In four probe technique, two of these probes contact outer side for the source of current. Other two probes

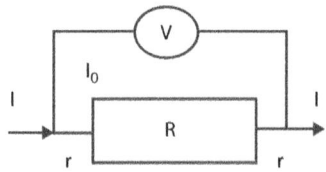

Figure 1.7 Schematic diagram of two probes electrical resistivity measurement [51].

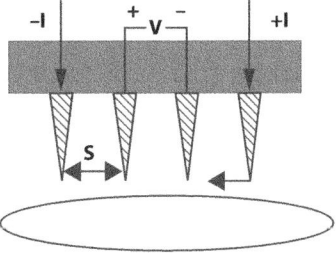

Figure 1.8 Schematic diagram of four probes electrical resistivity measurement [53].

are inner probes which measure voltage drop across the surface of materials. In four probe method, the voltage measured by the voltmeter using expression 1.6 as follows [53];

$$R = 2\pi S \left(\frac{V}{I}\right) \quad \text{...............(1.6)}$$

Van der Pauw method

Van der Pauw method [54] is adapted to measure sheet resistance of a material. Figure 1.9 shows schematic diagram of Van-der Pauw method for measuring electrical resistivity. Irregular shaped samples (shape and size) whose resistance cannot be measured two- and four-probe methods. In this method, two resistances are measured (R_A and R_B) using following equations 1.7, 1.8, 1.9 and 1.10 [54];

$$R_A = \left(\frac{V_{43}}{I_{12}}\right) \quad \text{...............(1.7)}$$

$$R_B = \left(\frac{V_{14}}{I_{23}}\right)$$

The sheet resistance is calculated by the van der Pauw relation

$$\left(e^{\frac{-\pi R_A}{R_S}}\right) + \left(e^{\frac{-\pi R_B}{R_S}}\right) = 1 \quad \text{...............(1.8)}$$

18 High Electrical Resistance Ceramics

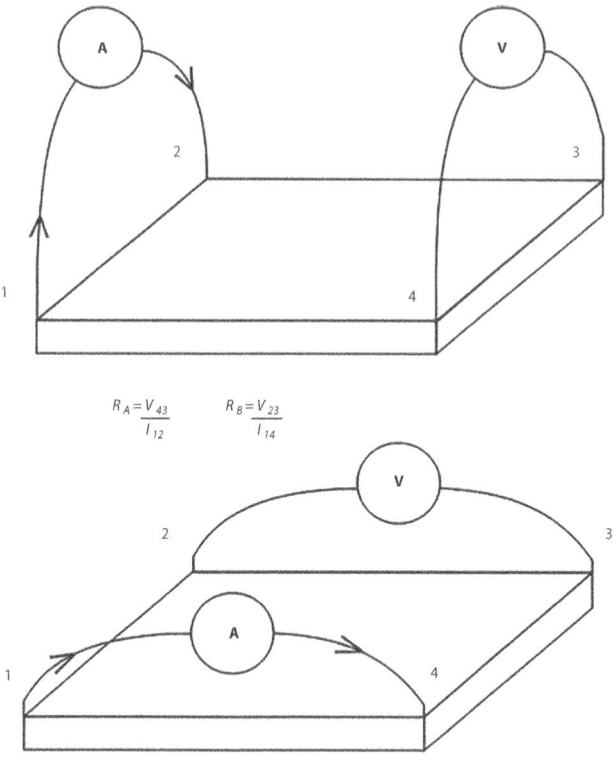

Figure 1.9 Schematic diagram of Van-der Pauw electrical resistivity measurement method [54].

Where, R_s is sheet resistance to be determined. If R_A and R_B are similar, then resistivity is given by,

$$\rho = \frac{\pi l}{\ln 2}\left(\frac{R_A + R_B}{2}\right) \quad \ldots (1.9)$$

Where, d is the thickness of the sample. If R_A and R_B are not similar then resistivity modifies as,

$$\rho = \frac{\pi l}{\ln 2}\left(\frac{R_A + R_B}{2}\right) f\left(\frac{R_A}{R_B}\right) \quad \ldots (1.10)$$

Where $f\left(\dfrac{R_A}{R_B}\right)$ is the function of the ratio $\left(\dfrac{R_A}{R_B}\right)$ only.

1.9 Resistance Temperature Detector (RTDs)

Resistance Temperature Detectors (RTDs) [55] are sensors. They are used to measure temperature by correlating resistance of RTD element with temperature. RTD elements comprise a length of fine coiled wire wrapped around a ceramic or glass core. This element is usually quite fragile. Thus, RTD elements are placed inside a sheathed probe, which is to be protected. RTD element is made from a pure material whose resistance has been recognized (at various temperatures). Some regularly used RTDs are Platinum (most popular and accurate), Nickel, Copper, Tungsten (rare) etc. [55].

1.9.1 Benefit of RTD

RTD [56] is a high accuracy, excellent stability, and repeatability temperature sensor. RTDs are relatively resistant to electrical noise. Consequently, it is suited for temperature measurement in industrial environments (i.e., around motors, generators and other high voltage equipment).

Gold and silver are rarely used as RTD elements because of lower resistivity and high cost. Tungsten is suited for high temperature applications. It has relatively high resistivity. Usually, RTDs are made of platinum, nickel, or nickel alloys. By comparison, nickel based wires are low cost. It is used in a limited temperature range. Nickel based RTD is non-linear and tends to drift with time. Platinum based RTDs are used for accuracy measurement.

1.10 Platinum Resistance Thermometer (PRTs)

Platinum Resistance thermometer [57] is also another kind of temperature sensor. It measures electrical resistance of platinum with variation of temperature. This type of sensor is being used in the place of thermocouples at < 600 °C, particularly for industrial applications. High temperature applications, it becomes difficult to operate because platinum is contaminated by impurities (from the metal sheath of the thermometer). PRTs require small current to pass through it and the resistance is determined with variation of temperatures. Resistance-temperature relationship of platinum is linear.

Different types of Pt based sensors are available in the market. These are Pt-100 and Pt-1000 sensors. The Pt-100 sensor has a resistance of 100 ohms at 0 °C and 138.4 ohms at 100 °C. Pt-1000 sensor has a resistance of 1000 ohms at 0 °C. Temperature dependent linear equation 1.11 is as follows [57];

$$R_t = R_0^* \left(1 + A \times t + B \times t^2 + C \times (t-100) \times t^3\right) \quad \ldots\ldots\ldots(1.11)$$

Here,
R_t is the resistance at temperature t
R_0 is the resistance at 0 °C
A = 3.9083×10⁻³,
B = -5.775 ×10⁻⁷
C = -4.183 ×10⁻¹² (below 0 °C)
C = 0 (above 0 °C).

1.11 Thermal Power Plant Wastes (i.e., Coal Ash) Management

Disposal of ashes is a great challenge to all stakeholders. Numerous initiatives are taken by the government & non-government organizations as well as research & development organizations to find out utilization of ashes i.e., wastes. Ashes are used in different sectors. Main sectors are construction i.e., Concrete production (as a substitute material) for Portland cement and sand, Embankments and other structural fills for road construction, brick production. Related activities are parallel going on to develop tiles ceramics, ceramic composites, flowable fill production, Waste stabilization & solidification. It is also used for cement clinkers production (as a substitute material for clay). Ashes are effectively used Mine reclamation, Stabilization of soft soils, Agricultural uses in soil amendment, fertilizer, cattle feeders, soil stabilization in stock feed yards, and agricultural stakes, cosmetics, toothpaste, kitchen countertops, floor and ceiling tiles, bowling balls, flotation devices, stucco, utensils, tool handles, picture frames, auto bodies and boat hulls, cellular concrete, geopolymers, roofing tiles, roofing granules, decking, fireplace mantles, cinder block, PVC pipe, Structural Insulated Panels, etc [58].

1.12 Literatures Survey on Thermal Power Plant Wastes-Based Ceramics

Looking at the compositions, Scientist/Researchers are putting their efforts to find new possibilities of consuming thermal power plant wastes. One of the possibilities is to use waste (20–50%) commercially to cement industry (as pozzolana/aggregate) [59]. Nowadays, fly ash is used mostly in building materials such as cement and concrete products, bricks, roadway and pavement, etc. Also, a small amount of fly ash is used as adsorbents (for flue gas), wastewater/soil improvement, and synthesis of zeolite [60], mullite [61], and glass-ceramics [62]. A particle size of fly ash varies between 1 and 100 μm. Major percentage composition of fly ash is SiO_2 and Al_2O_3. It is thought to be a low cost resource material for ceramic industries.

Major components present in fly ash are SiO_2 and Al_2O_3. In addition to two major components (SiO_2 and Al_2O_3), there is a minor phases of magnetite (Fe_3O_4)/hematite (Fe_2O_3) [60]. In Class F fly ashes, it contains trace amounts of CaO. In Class C fly ash, percentage of CaO is more in comparison to Class F fly ash [63]. Another important crystalline phase is mullite ($3Al_2O_3 \cdot 2SiO_2$). Usually, mullite is formed by the reaction between Al_2O_3- and SiO_2 [64]. Presence of mullite in fly ash containing mullite phase can be used for mullite-based products i.e., high-temperature applications. Mullites have a high temperature application (i.e., 1850 °C) since its melting point is high [65]. Another benefit is its stability either in oxidizing or reducing atmospheres [66]. It has moderate coefficient of thermal expansion [67], good thermal shock and spalling resistance, high shear modulus, low thermal conductivity, good resistance to acidic and neutral slags, and excellent general corrosion resistance.

Fly ash has a mullite phase but yet, the application of mullite-based refractory products is challenged. This is due to high glass content and high-temperature deformation [68]. Since fly ash contain ~60–70 wt% glass and therefore, it is highly vitreous bonds [69]. It follows by sigmoidal firing shrinkage kinetics. Slow rate softening enhances particle rearrangement and hence firing shrinkage decreases. Stoichiometry of mullite (i.e., $Al_2O_3 : SiO_2$) i.e., molar ratio of $Al_2O_3 : SiO_2$ is found to be show 3:2 to 2:1 [64, 67]. Single-crystal mullite is formed in needle-shaped or equiaxed grains. If $Al_2O_3 : SiO_2$ ratio is 3:2, mullite contents 'lower Al_2O_3. It results in more anisotropic grains [67]. Also, addition of oxides (e.g., Fe_2O_3) to the fly ash affects the morphology and hence, improves anisotropy [67]. Anisotropy is restrained by the addition of low-fluxing additives (e.g., Cr_2O_3) in high glass viscosities [70]. Mullite is an orthorhombic crystal unit cell and the preferred growth direction is [001] [71–73]. Reports are available on high-temperature axial thermal expansions, but interpolation of the most comprehensive set of data is 3:2 [74].

Generally, mullite-based materials are used in commercially available refractory products. Mullite (as mineral) is not acceptable as a raw material for fabricating refractory products [75]. Mullite usually is formed by the heat treatment process. Raw materials such as Aluminosilicates i.e., kaolinite ($Al_2O_3 \cdot 2SiO_2 \cdot 2H_2O$), andalusite/sillimanite/kyanite ($Al_2O_3\_SiO_2$), or Pyrophyllite ($4SiO_2Al_2O_3H_2O$) at temperature at 1200 °C is included. Such products comprise of glass-bonded aggregates [67]. Mullite-based refractory products are categorized on the basis of their Al_2O_3 contents [75, 76]. Creep resistance of single-phase mullite at 1400 °C is equivalent to creep resistance commercial mullites at 1000 °C [68]. This may happen due to the bonding between mullite grains and glass bonded refractories. Hence, thermal properties are dominated by the glassy matrix [77]. In refractory performance, thermal resistance is an essential factor, especially in metallurgical applications. Generally, Mullite-based refractories are used for linings of molten aluminium-alloy holding furnaces [78]. The molten metal wets and corrodes the inner-part of refractory. This leads to the reduction of siliceous component in mullite and thus Al_2O_3 is formed as a by-product. Such type of corrosion resistance is enhanced by the addition of non-wetting additives. The additives react with mullite to form complex oxides and that raises the resistance of the refractory surface [79–83]. Particularly, in mullite based products, presence of some transition metals (i.e., Ti, V, Fe, Zn, and Zr) reduces corrosion resistance of refractories [67], Elements present in mullite (Ni, Cr, etc) help to improve corrosion resistance. Stability of commercially mullite, present trend is to modify Al_2O_3:SiO_2 ratio in order to achieve porous mullite products. For porous mullite products, stoichiometric ratio of Al_2O_3 and SiO_2 to is nearer to the 3:2. Such type of porous ceramic production involves use of AlF_3, on SiO_2-rich components. AlF_3 eradicates the glass and makes its porosity. Heating time is about 4 h. Below this time, whiskers are formed. Reports are not available on long-term firing shrinkages [84].

Nowadays, especially, huge amounts of fly ash is consumed in the construction areas as building materials such as cement/concrete based products, fly ash based bricks, fly ash based roadway, and fly ash based pavement materials [85–89]. Other sectors such as adsorbents (for flue gas), wastewater/soil improvement, and utilized for synthesis of zeolite, etc. are used for small quantities of fly ash [90–92]. Looking to the properties (such as physical and chemical) and chemical composition of fly ash, it is considered as a low cost material resource and is employed in ceramic industries.

Other wastes, red mud is obtained or discharged during production of alumina from bauxite ores by Bayer Process. Red mud can be divided into two types. One of red mud is obtained from Bayer Process and other type of red mud is produced by calcination bauxite. Physical, chemical and mineralogical compositions fluctuate during aluminium preparation processes [93]. Major constituents of red mud are SiO_2, Al_2O_3, Fe_2O_3 and CaO. In the present scenario, researchers have explored the utilization of red mud in the area of heavy metal immobilization, production of Portland cement clinker, coagulant, adsorbent and catalyst for environmentally benign processes, etc. [94–96].

Red mud is used as raw materials to prepare thermal storage ceramics, which have high thermal conductivity and high heat storage density [97]. For obtaining high thermal conductivity and high heat storage density, fly ash powders are added to red mud during preparation of thermal storage ceramics and thereby, improving thermal shock resistance.

Ceramic matrix composites (CMCs) have better durability and strength properties. These may be used for aero and nuclear turbine hot-sections. In CMC systems, fiber-reinforced silicon carbide (SiC) composite is prepared by melt-infiltrated (MI) process. These composites materials can withstand high temperature upto 1315 °C [98–100]. Such composite material which prepared by MI process are ascribed to improve densification, competing with SiC_f/SiC processed by CMC processing methods (i.e., polymer infiltration and pyrolysis (PIP) and chemical vapor infiltration (CVI)). Decrease in matrix porosity supports a decreased environmental attack on the reinforcing fibers and increases the load carrying capability of the matrix. It is difficult task to find out and recognize damage mechanisms in their operating environments. Constituents of CMCs are semiconducting nature. Electrical resistance (ER) measurements have supported to detect damage accumulation in the form of transverse matrix cracking, in both CVI [101] and MI-CVI SiC_f/SiC CMCs [102] at room temperature under tensile loading [102]. Recently, ER (Electrical resistance) is used to characterize high temperature creep behaviors of MI composites under furnace-heated or laser-heated in thermal gradient conditions [103]. Thermomechanical test results on SiC_f/SiC CMCs in oxidizing environments are correlated with ER response [104]. So, Electrical resistance measurement helps to monitor longevity of composite. Also, Electrical resistance is prospective to understand the contributions of thermal/mechanical loading and environmental effects. Electrical properties of MI-CVI SiC_f/SiC composites under thermal loading conditions are well established. A few reports are available on temperature dependence electrical resistivity on different SiC fiber reinforced in polycrystalline SiC matrix [105, 106].

ER behaviour of MI composites (to thermal loading) is sometime differs considerably. This is due to nature of siliconized silicon carbide i.e., Si-SiC matrix formed during molten silicon infiltration method. Excess free silicon (Si) is kept in the matrix after completion of this process. It will tend to show more conductive if compare with pure CVI matrix. Silicon SiC_f/SiC based composite is prepared by a melt-infiltrated method using different reinforcing fiber (Hi-Nicalon Type S, Tyranno SA and ZMI). Temperature dependent electrical resistivity of silicon SiC_f/SiC based composite are measured.

Ceramics are solid compounds, which entail metallic and nonmetallic elements. It is formed by the application of pressure or load followed by heat-treatment [107]. Commonly, ceramics are: hard, wear-resistant, brittle, refractory, electrical insulators, oxidation resistance, thermal shock prone, chemically stable, etc [108]. Recently, ceramic materials are used in every sector. Excellent properties of ceramics are due to strong bonding between the atoms and way of packing. In ceramics both (covalent and ionic) type of bonding are feasible. In comparison to metals on electrical resistivity, ceramics have very high electrical resistivity because of their strong ionic-covalent bonding or non-availability of free electrons [109].

Best of our knowledge, ceramic are dielectrics and have high electrical resistivity. But It supports electrostatic field. Electrical resistivity of ceramics is varied with the frequency (of the applied field) and temperature [110]. Charge transport phenomenon is frequency dependent, whereas thermal energy provides the activation energy for the charge migration. In general, ceramics have high value of dielectric strength and dielectric constant. Ceramic based dielectric materials have different uses such as capacitors, insulators and resistors for electronic devices [111].

Because of high electrical resistivity value, dielectrics show electrical insulating behaviour and can be polarized by the application of electric field. All types of dielectrics are insulators but not vice-versa (i.e., all the insulators are not dielectrics). Dielectrics are two types. One is non-polar and other one is polar. By the application of an electric field to dielectrics, displacement of center of positive and center of negative charge occurs. It produces a dipole moment. By this way, it stores energy. In dielectrics, elementary dipoles interact with each other under certain thermodynamic circumstances.

Fore going discussion is made on waste materials from thermal power plants. In addition, ceramics, insulators, dielectrics in general and fly ash/pond ash in particular are briefly discussed. Possibilities of making usefully ceramics, insulators, and dielectrics have reviewed.

1.13 Conclusions

Despite shifting the energy generation modes from thermal to other renewable sources, the dependence on coal is expected to be significant in the next decade. This will result generation of substantial amount of fly ash/pond ash/bottom ash. It is essential to explore possibilities to use thermal power plant wastes. This chapter displays potential use of ashes in various construction sectors such as cement, concrete, bricks, and blocks (using a simple framework). Some recent applications include tiles panels and composite materials. Present research suggests that stakeholders must work together to upgrade the existing policies required for the successful commercialization of these products in the market. This study forms a base for future research in which barriers like fly ash/pond ash/bottom ash be overcome. Fruitful utilization of this material is investigated comprehensively. Therefore, a plan of action has to chalk out fruitful utilization of these wastes and to develop cheaper and useful materials.

Acknowledgements

First author would like to thank Prof. Munesh Chandra Adhikary, PG Council Chairman, Fokir Mohan University, for his invaluable guidance, advice, and constant inspiration throughout the entire program. First author thanks Mr. Mukteswar Mohapatra in Fokir Mohan University for their support. The authors convey their sincere thanks to GIET, University Gunupur, Rayagada, Odisha, India for providing Lab facilities to do the research work. The authors also thank the CRF, IIT Kharagpur for providing their testing facilities.

References

1. Birol, Fatih; Malpass, David. "It's critical to tackle coal emissions – Analysis". *International Energy Agency*. Retrieved 9 October 2021.
2. M. Cropper, R. Cui, S. Guttikunda, N. Hultman, P. Jawahar, Y. Park, X. Yao, and X.-P. Song, (2021), the Mortality Impacts of Current and Planned Coal-Fired Power Plants in India, *Proceedings of the National Academy of Sciences*, Vol. 118.
3. A. Z. AL Shaqsi, K. Sopian, and A. Al-Hinai, (2020), Review of Energy Storage Services, Applications, Limitations, and Benefits, *Energy Reports*, Vol. 6, pp. 288-306 https://doi.org/10.1016/j.egyr.2020.07.028.
4. www.flyash.com/ and www.iflyash.com/
5. A Wardhono, (2017), Comparison Study of Class F and Class C Fly Ashes as Cement Replacement Material on Strength Development of Non-Cement Mortar, IOP Conference Series: Materials Science and Engineering, Vol. 288, *The 2nd Annual Applied Science and Engineering Conference (AASEC 2017)* 24 August 2017, Bandung, Indonesia.
6. M.K. Panigrahi, (2021), Investigation of Structural, Morphological, Resistivity of Novel Electrical Insulator: Industrial Wastes, *Bull. Sci. Res.*, Vol. 3, pp. 51-58.

7. M. H. Abdullah, A. S. A. Rashid, U. H. M. Anuar, A. Marto, and R. Abuelgasim, (2019), Bottom ash utilization: A review on engineering applications and environmental aspects GEOTROPIKA 2019, *IOP Conf. Series: Materials Science and Engineering*, Vol. 527, 012006.
8. Nurul izzati raihan ramzi hannan, Shahiron Shahidan, Mohamad Zulkhairi Maarof, and Noorwirdawati Ali, (2016), Physical and Chemical Properties of Coal Bottom Ash (CBA) from Tanjung Bin Power Plant, *IOP Conference Series Materials Science and Engineering* 160(1):012056.
10. T. Janardhanan and R. Venkatasubramani, (2015), Properties of Foundry Sand, Ground Granulated Blast Furnace Slag and Bottom Ash Based Geopolymers under Ambient Conditions, *Periodica Polytechnica Civil Engineering*, Vol. 60, pp.1-10.
11. M. K. Panigrahi, R. R. Dash, R. I. Ganguly, (2016), Optimization Mechanical Properties of Pond Ash Geopolymer: As Construction Material, *National Conference on Advanced Engineering Materials-2016* (23-24th July 2016).
12. M. K. Panigrahi, P. K. Rana, A. K. Pradhan, P. K. Rout, A. K. Samal, S. Gupta and Mv B. Kumar, Production of Geopolymer based Construction Material from Pond Ash: An Industrial Waste, *19th International Conference Non-Ferrous Metal-2015* (9,10[th] July), pp. 190-200.
13. M. Hashan, M. F. Howladar, L. N. Jahan, and P. K. Deb, (2013), Ash Content and Its Relevance with the Coal Grade and Environment in Bangladesh, *International Journal of Scientific & Engineering Research*, Vol. 4, pp. 669-676.
14. Coal Fly Ash - User Guidelines for Waste and Byproduct Materials in Pavement Construction- Material Description-FHWA-RD-97-148.
15. R. Sokolar and M. Nguyen, (2018), The Fly Ash of Class C for Ceramic Technology, *IOP Conf. Series: Materials Science and Engineering*, Vol. 385, pp. 012053-xxx.
16. A. Wardhono, (2017), Comparison Study of Class F and Class C Fly Ashes as Cement Replacement Material on Strength Development of Non-Cement Mortar, IOP Conference Series: Materials Science and Engineering, Vol. 288 *The 2nd Annual Applied Science and Engineering Conference (AASEC 2017)* 24 August 2017, Bandung, Indonesia.
17. American Coal Ash Association, (1996), Coal Combustion Product-Production and Use. Alexandria, Virginia, 1997.
18. T.-C. Ke, and C. W. Lovell, (1992), Corrosively of Indiana Bottom Ash, Transportation Research Record Number-1345, Transportation Research Board, Washington, DC, 1992.
19. K. L. Moulton, (1973), Bottom Ash and Boiler Slag, Proceedings of the Third International Ash Utilization Symposium, U.S. Bureau of Mines, Information Circular No. 8640, Washington, DC, 1973.
20. J. R. Dungca and J. A. L. Jao, (2017), Strength and Permeability Characteristics of Road Base Materials Blended with Fly Ash and Bottom Ash, *International Journal of GEOMATE*, Vol. 12, pp. 9-15.

21. N. I. R. Ramzi, S. Shahidan, M. Z. Maarof, N. Ali, (2016), Physical and Chemical Properties of Coal Bottom Ash (CBA) from Tanjung Bin Power Plant, *IOP Conf. Series: Materials Science and Engineering*, Vol. 160, pp. 1-10.
22. U.S. Environmental Protection Agency (EPA), Washington, D.C. "Hazardous and Solid Waste Management System; Identification and Listing of Special Wastes; Disposal of Coal Combustion Residuals from Electric Utilities, Proposed rule. *Federal Register, 75 FR 35130*, June 21, 2010.
23. Effluent Limitations Guidelines and Standards for the Steam Electric Power Generating Point Source Category, *EPA*, 2018-11-30 https://www.epa.gov/eg/steam-electric-power-generating-effluent-guidelines-2015-final-rule.
24. Brooke, Nelson, (June 5, 2019). "New Interactive Maps of Groundwater Pollution Reveal Threats Posed by Alabama Power Coal Ash Pits, Black Warrior River keeper, Birmingham, AL https://blackwarriorriver.org/new-coal-ash-pollution-maps/
25. Springer, Patrick, (2019), Report: Unsafe Coal Ash Contamination Found in North Dakota Ground Water, Bismarck Tribune. Bismarck, ND https://bismarcktribune.com/news/state-and-regional/report-unsafe-coal-ash-contamination-found-in-north-dakota-groundwater/article_ccbda15d-5489-55f0-a756-84ebb7dd062f.html.
26. T. Tosheff, (2019), York: Brunner Island Power Plant Owners Agree to $1M Penalty, Coal Ash Cleanup, Harrisburg, PA: *ABC27 News* https://www.abc27.com/news/local/york/brunner-island-power-plant-owners-agree-to-1m-penalty-coal-ash-cleanup.
27. P. Ghosh and S. Goel, (2014), Physical and Chemical Characterization of Pond Ash, *International Journal of Environmental Research and Development*. Vol. 4, pp. 129-134.
28. A. Bhatt, S. Priyadarshini, A.A. Mohanakrishnan, A. Abri, M. Sattler, S. Techapaphawit, (2019), Physical, Chemical, and Geotechnical Properties of Coal Fly Ash: A Global Review Author Links Open Overlay Panel, *Case Studies in Construction Materials*, Vol. 11, pp. 00263-xxx https://doi.org/10.1016/j.cscm.2019.e00263.
29. S.M. Rao and I.P. Acharya, (2014), Synthesis and Characterization of Fly Ash Geopolymer Sand, *Journal of Materials in Civil Engineering ASCE*, Vol. 26, pp.912-917.
30. J. Feng, J. Sun, and P. Yan, (2018), The Influence of Ground Fly Ash on Cement Hydration and Mechanical Property of Mortar, *Advances in Civil Engineering*, Vol. 2018, pp. 7.
31. K. ZabielskaAdamska, (2020), Hydraulic Conductivity of Fly Ash As a Barrier Material: Some Problems in Determination, *Environmental Earth Sciences*, Vol. 79, pp.321-333.
32. H. Li, G. Liu, and Y. Cao, (2014), Content and Distribution of Trace Elements and Polycyclic Aromatic Hydrocarbons in Fly Ash from a Coal-Fired CHP Plant, *Aerosol and Air Quality Research*, Vol. 14, pp. 1179-1188.

33. G. A. Leonards and B. Bailey, (1982), Pulverized Coal Ash as Structural Fill, *Journal of Geotechnological Engineering, ASCE*, Vol. 108, pp.517-531.
34. Coal Physical Testing, SGS Minerals Services-T3 SGS 527, 10-2013. Email us at minerals@sgs.com www.sgs.com/coal
35. A. Sridharan and K. Prakash, (2000), Classification Procedures for Expansive Soils, *Proceedings of the Institution of Civil Engineers Geotechnical Engineering*, Vol. 143, pp. 235-240.
36. W.R. Roy and P.M. Berger, (2011), Geochemical Controls of Coal Fly Ash Leachate pH, *Coal Combustion and Gasification Products*, Vol. 3 pp.63-66.
37. P. Trinh, B. Yuko, O. Kenichiro, and N. K. Kawai, (2015), A Study on Pozzolanic Reaction of Fly Ash Cement Paste Activated by an Injection of Alkali Solution, *Construction and Building Materials*, Vol. 94, pp. 28-34.
38. N. Pandian, A. Sridharan, and S. Srinivas, (2001), Compaction Behaviour of Indian coal Ashes, *Proceedings of the Institution of Civil Engineers Ground Improvement*, Vol. 5, pp. 13-22.
39. T. Xie and T. Ozbakkaloglu, (2015), No Access Influence of Coal Ash Properties on Compressive Behaviour of FA- and BA-based GPC, *Magazine of Concrete Research*, Vol. 67, pp. 1301-1314.
40. S. R. Kaniraj and V. Gayathri, (2004), Permeability and Consolidation Characteristics of Compacted Fly Ash, *Journal Energy Engineering*, Vol. 130, pp. 18-43.
41. M. R. Hajarnavis and A. D. Bhide, (2003), Leaching Behaviour of Coal-Ash: A Case Study, *Indian J Environ Health*, Vol. 45, pp. 293-8.
42. https://ceramics.org/about/what-are-engineered-ceramics-and-glass/structure-and-properties-of-ceramics.
43. https://electrical-engineering-portal.com/ceramic-porcelain-and-glass-insulators
44. S. Sahoo, U. Dash, S. K. S. Parashar, and S. M. Ali, (2013), Frequency and Ttemperature Dependent Electrical Characteristics of $CaTiO_3$ Nano-Ceramic Prepared by High-Energy Ball Milling, *Journal of Advanced Ceramics*, Vol. 2, pp. 291-300.
45. https://ceramics.org/about/what-are-engineered-ceramics-and-glass/ceramics-and-glass-in-electrical-and-electronic-applications
46. G. S. Ohm, German physicist who discovered the law, named after him, which states that the current flow through a conductor is directly proportional to the potential difference (voltage) and inversely proportional to the resistance.
47. https://www.elprocus.com/what-is-a-resistor-construction-circuit-diagram-and-applications/
48. https://en.wikipedia.org/wiki/Insulator_(electricity).
49. https://ns.ph.liv.ac.uk/~ajb/radiometrics/glossary/insulator.html.
50. https://openpress.usask.ca/physics155/chapter/5-3-resistivity-and-resistance/.
51. Y. Singh, (2013), Electrical Resistivity Measurements: A Review, International Conference on Ceramics, Bikaner, India, *International Journal of Modern*

Physics: Conference Series, Vol. 22, pp. 745-756 http//doi/abs/10.1142/ S2010194513010970.
52. Y. Singh, (2013), Electrical Resistivity Measurements: a Review, *International Journal of Modern Physics Conference Series*, Vol. 22, pp. 745-756.
53. https://www.nitsri.ac.in/Department/PHYSICS/1_Resistivity_by_Four_Probe_Method.
54. Van der Pauw method, Wikipedia, the free encyclopedia
55. https://www.te.com/usa-en/industries/sensor-solutions/insights/understanding-rtds.html#:~:text=An%20RTD%20(Resistance%20Temperature%20Detector,RTD%20is%20a%20passive%20device.
56. https://www.polytechnichub.com/advantages-disadvantages-rtd-resistance-temperature-detector/
57. I. Yang, Suherlan, K. S. Gam and Y.-G. Kim, (2015), Interpolating equation of Industrial Platinum Resistance Thermometers in the Temperature Range Between 0°C and 500°C, *Measurement Science and Technology*, Vol. 26, pp. 035104-7.
58. CHAPTER 6 -INDUSTRIAL SOLID WASTE, National Waste Management Council- Ministry of Environment & Forests-1990/1999.
59. G. M. SadiqulIslam, M. H. Rahman, and Nayem Kazi, (2017), Waste Glass Powder as Partial Replacement of Cement for Sustainable Concrete Practice, *International Journal of Sustainable Built Environment*, Vol. 6, pp. 37-44 https://doi.org/10.1016/j.ijsbe.2016.10.005.
60. K. Ojha, N. C. Pradhan and A. Nath Samanta, (2004), Zeolite from Fly Ash: Synthesis and Characterization, *Bulletin of Materials Science*, Vol. 27, pp. 555–564.
61. T. F. Choo, Mohamad A., Mohd S., K. Y. Kok, and K. A. Matori, (2019), A Review on Synthesis of Mullite Ceramics from Industrial Wastes, *Recycling*, Vol. 4, pp. 1-12.
62. J. Luan, A. Li, T. Su, and X. Cui, (2009), Synthesis of Nucleated Glass-Ceramics using Oil Shale Fly Ash, *Journal of Hazardous Materials*, Vol. 173, pp. 427-32.
63. S. Shirkhanloo, M. Najafi, V. Kaushal, and M. Rajabi, (2021), A Comparative Study on the Effect of Class C and Class F Fly Ashes on Geotechnical Properties of High-Plasticity Clay, *Civil Engineering*, Vol. 2, pp. 1009-1018 https://doi.org/10.3390/civileng2040054.
64. B. Saruhan, W. Albers, H. Schneider, and W. A. Kaysser, (1996), Reaction and Sintering Mechanisms of Mullite in the Systems Cristobalite/α-Al_2O_3 and Amorphous SiO_2/α-Al_2O_3, *Journal of the European Ceramic Society*, Vol. 16, pp. 1075-1081 https://doi.org/10.1016/0955-2219(96)00023-4.
65. R. Torricelli's, J. M. C. Jose, S. Moya, M. J. Reece, C. K. L. Davies, C. Olagnond and G Fantozzid, (1999), Suitability of Mullite for High Temperature Applications, *Journal of the European Ceramic Society*, Vol. 19, pp 2519-2527 https://www.researchgate.net/publication/248453468.

66. J. Anggono, (2005), Mullite Ceramics: Its Properties, Structure, and Synthesis, *Jurnal Teknik Mesin*, Vol. 7, pp. 1-10.
67. V. Viswabaskaran, F. D. Gnanam, and M. Balasubramanian, (2004), Mullite from Clay-Reactive Alumina for Insulating Substrate Application, *Applied Clay Science*, Vol. 25, pp. 29-35 https://doi.org/10.1016/j.clay.2003.08.001.
68. D. Zemánek, K. Lang, L. Tvrdík, D. Všianský, L. Nevřivová, P. Kovář, L. Keršnerová, and K. Dvořák, (2021), Development and Properties of New Mullite Based Refractory Grog, *Materials*, Vol. 14, pp. 779-15.
69. C. R. Ward and D. H. French, (2006), Determination of Glass Content and Estimation of Glass Composition in Fly Ash using Quantitative X-Ray Diffractometry, *Fuel*, Vol. 85, pp. 2268-2277.
70. B. M. Kim, Y. K. Cho, S.-Y. Yoon, R. Stevens, and H. C. Park, (2009), Mullite Whiskers Derived from Kaolin, *Ceramics International*, Vol. 35, pp. 579-583.
71. H. Schneider, R. Fischer, and J. Schreuer, (2015), Mullite: Crystal Structure and Related Properties, *Journal of the American Ceramic Society*, Vol. 98, pp. 2948-2967.
72. R. Fischer and H. Schneider, (2000), Crystal Structure of Cr-Mullite, American Mineralogist, Vol. 85, pp. 1175-1179.
73. https://application.wiley-vch.de/books/sample/3527309748_c01.pdf
74. A. Abdullayev, F. Zemke, A. Gurlo, and M. F. Bekheet, (2020), Low-Temperature Fluoride-Assisted Synthesis of Mullite Whiskers, *RSC Advances*, Vol. 10, pp. 31180-31186.
75. P. Koshy, N. Ho, V. Zhong, L. Schreck, S. A. Koszo, E. J. Severin, and C. C. Sorrell, (2021), Fly Ash Utilization in Mullite Fabrication: Development of Novel Percolated Mullite, *Minerals*, Vol. 11, pp. 1-16 https://doi.org/10.3390/min11010084.
76. K. Dana, S. Sinha Mahapatra, H. S. Tripathi, and A. Ghosh, (2014), Refractories of Alumina-Silica System, Transactions-Indian Ceramic Society, Vol. 73, pp. 1-13.
77. S. C. Carniglia and G. L. Barna, (1992), Handbook of Industrial Refractories Technology: Principles, Types, Properties and Applications; Noyes Publications: Park Ridge, NJ, USA.
78. F. Barandehfard, J. Aluha, A. Hekmat-Ardakan, and F. Gitzhofer, (2020), Improving Corrosion Resistance of Aluminosilicate Refractories towards Molten Al-Mg Alloy Using Non-Wetting Additives: A Short Review, *Materials*, Vol. 13, pp. 4078-26.
79. C. Sadik, I. E. El Amrani, and A. Albizane, (2014), Recent Advances in Silica-Alumina Refractory: A Review, *Journal of Asian Ceramic Societies*, Vol. 2, pp. 83-96.
80. S. Afshar and C. Allaire, (2000), The Corrosion Kinetics of Refractory by Molten Aluminium, *JOM*, pp. 43-46.
81. R. N. Nandy and R. K. Jogai, (2013), Selection of Proper Refractory Materials for Energy Saving in Aluminium Melting And Holding Furnaces, *International Journal of Metallurgical Engineering*, 1, 117–121.

82. S. Afshar, and C. Allaire, (2004), Protection of Aluminosilicate Aggregates Against Corrosion by Molten Aluminum, *Report*, Vol. 1–13, pp. 279-290.
83. L. F. Hou, Y. H. Wei, Y. G. Li, B. S. Liu, H. Y. Du, and C. L. Guo, (2013), Erosion Process Analysis of Die-Casting Inserts for Magnesium Alloy Components, *Engineering Failure Analysis*, Vol. 33, pp. 457-464.
84. Z. Yang, F. Yang, S. Zhao, K. Li, J. Chen, Z. Fei, and G. Chen, (2021), *In-situ* Growth of Mullite Whiskers and their Effect on the Microstructure and Properties of Porous Mullite Ceramics with an Open/Closed Pore Structure, *Journal of the European Ceramic Society*, Vol. 41, pp. xxx-xxx.
85. N. Ghazali, K. Muthusamy, and S. Wan Ahmad, (2019), Utilization of Fly Ash in Construction, *IOP Conf. Series: Materials Science and Engineering*, Vol. 601, pp. 1-9.
86. W. M. Wan Ibrahim, M. M. Al Bakri Abdullah, A. V. Sandu, K. Hussin, I. G. Sandu, K. N. Ismail, A. Abdul Kadir, Md. Binhussain, (2014), Processing and Characterization of Fly Ash-Based Geopolymer Bricks, *REVISTA DE CHIMIE (Bucharest)*, Vol. 65, pp. 1340-1345.
87. C. Feronea, F. Colangeloa, R. Cioffia, F. Montagnarob, L. Santorob, (2011), Mechanical Performances of Weathered Coal Fly Ash based Geopolymer Bricks, International Conference on Green Buildings and Sustainable Cities, *Procedia Engineering*, Vol. 21, pp. 745-752.
88. M. Yang, M. ASCE, Shree R. Paudel, and Z. Jerry Gao, (2012), Snow-Proof Roadways Using Steel Fiber–Reinforced Fly Ash Geopolymer Mortar–Concrete, *Journal of Materials in Civil Engineering*, Vol. 33, pp. 04020444-11 https://ascelibrary.org/doi/epdf/10.1061/%28ASCE%29MT.1943-5533.0003537.
89. T. Poltue, A. Suddeepong, S. Horpibulsuk, W. Samingthong, and A. Arulrajah, (2020), Strength Development of Recycled Concrete Aggregate Stabilized with Fly Ash-Rice Husk Ash based Geopolymer as Pavement Base Material, *Road Materials and Pavement Design*, Vol. 21, pp. 2344-2355.
90. Saakshy, K. Singh, A. B. Gupta, and A. K. Sharma, (2016), Fly Ash as Low Cost Adsorbent for Treatment of Effluent of Handmade Paper Industry-Kinetic and Modelling Studies for Direct Black Dye, *Journal of Cleaner Production*, Vol. 112, pp. 1227-1240.
91. U. P. Nawagamuwa and N. Wijesooriya, (2018), Use of Flyash to Improve Soil Properties of Drinking Water Treatment Sludge, *International Journal of Geo-Engineering*, Vol. 9, pp. 1-8.
92. S Subhapriya and P Gomathipriya, (2018), Synthesis and Characterization of Zeolite X from Coal Fly Ash: A Study on Anticancer Activity, *Materials Research Express*, Vol. 5, pp. 085401-12.
93. N. C. G. Silveira, M. L. F. Martins, A. C. S. Bezerra, and F. G. S. Araújo, (2021), Red Mud from the Aluminium Industry: Production, Characteristics, and Alternative Applications in Construction Materials-A Review, *Sustainability*, Vol. 13, pp. 12741-21 https://doi.org/10.3390/su132212741.

94. Y. Hua, K. V. Heal, and W. Friesl-Hanl, (2016), the Use of Red Mud as an Immobiliser for Metal/Metalloid-Contaminated Soil: A Review, *Journal of Hazardous Materials*, Vol. 325,pp. 17-30.
95. A. Zhao, Y. Liu, T.-A. Zhang, X. He, X. Ye, and M. Zeng, (2022), Preparation and Characterization of Portland Cement Clinker from Sulfuric Acid Leaching Residue of Coal Fly Ash, *Materials Research Express*, Vol. 9, pp. 035202-11 https://doi.org/10.1088/2053-1591/ac4e3b.
96. U. O. Aigbe, K. E. Ukhurebor,, R. B. Onyancha, O. A. Osibote, H. Darmokoesoemo and H. S. Kusuma, (2021), Fly ash-based Adsorbent for Adsorption of Heavy Metals and Dyes from Aqueous Solution: A Review, *Journal of Materials Research and Technology*, Vol. 14, pp. 2751-2774 https://doi.org/10.1016/j.jmrt.2021.07.140.
97. J. LI, B. Yan, Z. Cheng, and T. Deng, (2019), Utilization of Fly Ash and Red Mud for Thermal Storage Ceramics, *Journal of the Ceramic Society of Japan*, Vol. 127, pp. 931-938.
98. F. W. Zok, (2016), Ceramic-Matrix Composites Enable Revolutionary Gains in Turbine Engine Efficiency, *American Ceramic Society Bulletin*, Vol. 95, pp. 22-28.
99. M. Verrill, R. C. Robinson, A. M. Calomino, and D. J. Thomas, (2004), Ceramics Matrix Composite Vane Sub-Element Testing in a Gas Turbine Environment, ASME paper, GT200453970, *Proceeding of ASME Turbo Expo 2004*, June 14-17, 2004 Vienna, Austria.
100. G. N. Morscher, and V. V. Pujar, (2009), Design Guidelines for In-Plane Mechanical Properties of SiC Fiber-Reinforced Melt-Infiltrated SiC Composites, *International Journal of Applied Ceramic Technology*, Vol. 6, pp. 151-163 https://doi.org/10.1111/j.1744-7402.2008.02331.x.
101. G. Morscher, C. E. Smith, E. Maillet, C. Baker, and R. Mansour, (2014), Electrical Resistance Monitoring of Damage and Crack Growth in Advanced SiC-based Ceramic Composites, *American Ceramic Society Bulletin*, Vol. 93, pp. 28-31.
102. G. N. Morscher and V. V. Pujar, (2006), Creep and Stress–Strain Behavior After Creep for SiC Fiber Reinforced, Melt-Infiltrated SiC Matrix Composites, *Journal of American Ceramic Society*, Vol. 89, pp. 1652-1658.
103. T. Whitlow, J. Pitz, J. Pierce, S. Hawkins, A. Samuel, K. Kollins, G. Jefferson, E. Jones, J. Vernon, and C. Przybyla, (2019), Thermal-Mechanical Behavior of a SiC/SiC CMC Subjected to Laser Heating, *Composite Structures*, Vol. 210, pp. 179-188 https://doi.org/10.1016/j.compstruct.2018.11.046.
104. Ragavendra P. Panakarajupally, Michael J. Presby, Manigandan K, George G. Chase, Jianyu Zhou, and Greg Morscher, (2019), Thermomechanical Characterization of SiC/SiC Ceramic Matrix Composites in a Combustion Facility, *Ceramics*, Vol. 2, pp. 407-425.
105. R. P. Panakarajupally, M. K. Gregory, and N.Morscher, (2021), Tension-Tension Fatigue Behavior of a Melt-Infiltrated SiC/SiC Ceramic Matrix Composites in

a Combustion Environment, *Journal of the European Ceramic Society*, Vol. 41, pp. 3094-3107 https://doi.org/10.1016/j.jeurceramsoc.2020.10.007.
106. A.R. Raffray, R. Jones, G. Aiello, M. Billone, L. Giancarli, H. Golfier, A. Hasegawa, Y. Katoh, A. Kohyama, S. Nishio, B. Riccardi, and M.S. Tillack, (2001), Design and Material Issues for High Performance SiCf/SiC-based Fusion Power Cores, *Fusion Engineering and Design*, Vol. 55, pp. 55-95.
107. T. Majeed, Mohd A. Wahid, N. Sharma, (2017), Ceramic Materials: Processing, Joining and Applications, IJCRT, *International Conference Proceeding ICCCT* Dec 2017, ISSN: 2320-2882 IJCRTICCC025, pp. 166-171.
108. F. Cardarelli, Ceramics, Refractories, and Glasses, (2008), Materials Handbook, Springer, London, Print, pp 593-689.
109. H.-P. Martin, (2014), Conductive Ceramics as Electrical Materials at High Temperatures, Annual Report 2014/15, pp. 38-xxx.
110. N. V. Gorshkov, V. Goffman, M. A. Vikulova, E. V. Tretyachenko, D. S. Kovaleva, and Alexander V. Gorokhovsky, (2018), Temperature-Dependence of Electrical Properties for the Ceramic Composites based on Potassium Polytitanates of Different Chemical Composition, *Journal of Electroceramics*, Vol. 40, pp.1-10.
111. W. Luo, S. Yan, and J. Zhou, (2022), Ceramic-based Dielectric Meta-Materials, *Interdisciplinary Materials*, Vol. 1, pp. 11-27.

2

Ceramic Production Methods and Basic Characterization Techniques

Muktikanta Panigrahi[1*], Ratan Indu Ganguly[2] and Radha Raman Dash[3]

[1]*Department of Materials Science, Maharaja Sriram Chandra BhanjaDeo University, Balasore, Odisha, India*
[2]*Department of Metallurgical Engineering, National Institute of Technology, Raurkela, Odisha, India*
[3]*CSIR-National Metallurgical Laboratory, Jamshedpur, Jharkhand, India*

Abstract

Extensive review made on production of Ceramics. Literature survey has been made on production of ceramic and its products using waste materials such silica fumes, low- & high-carbon fly ash, redmud, high- & low-carbon blast furnace slag, rice-husk ash, zeolites, argon oxygen decarburization (AOD) slag, bottom ash, pond ash, etc. Different production methods adopted are curing, sol-gel, deep mixing, hydrothermal synthesis and low-temperature calcination methods, mechanical activation (MA), blended mineral admixtures, an alkaline solution of thermally activated silica-alumina bearing mineral with additives such as calcium carbonate, microwave radiation, sintering process. Plenty of data are cited to show effect of variables on production of ceramic. Influences of different variables on production of ceramics are described.

Keywords: Industrial waste, pond ash, slag, ceramics, synthesis, characterization techniques

2.1 Introduction

Ceramic is a Greek word derived from karamus [1]. Earlier it was known as "pottery" because pots or vessels are made from ceramics. Ceramics is an inorganic, nonmetallic or metalloid solid compound, which has high

---

Corresponding author: muktikanta2@gmail.com

melting temperature and high hardness value. In the present scenario, ceramic has more roomy meaning i.e., glazed, smooth, colored surfaces, decreasing porosity through use of glassy, amorphous ceramic coatings on top of the crystalline ceramic surface. Ceramic has some unique characteristics i.e., high melting point, good chemical inertness, brittleness, high-temperature stability, heat and electrical insulation ability [2]. Hence, ceramic has huge potential engineering applications in modern society. It is believed that ceramic will be more strongly demanded in the near future [3]. In day-to-day life, bricks, glass, tiles, table wares, sanitary wares are required for domestic purpose. Magnesite, chrome, dolomite, zircon, quartz, and little or no clay are used as raw materials for refractory preparations. Also, ceramics are widely employed in advanced fields viz., electrical, electronics, nuclear power, and structural engineering. Nitrides or carbides are used for making heating elements, abrasives and structural components, construction materials, whereas zirconia, beryllia, and thoria are used for preparing advanced refractories. Rutile phase of ceramic is used for producing ferroelectric materials and uranium oxide is used as a nuclear fuel element [4, 5]. Because of the huge consumption of naturally available ceramic raw materials by ceramic industries, there is a shortage of these natural resources. Therefore, scientific communities are putting their efforts to discover supernumerary ceramic ingredients and protect the ecosystem.

Need for ceramic material is growing exponentially with increasing population of the world. Therefore, it is essential to utilize waste from different industries, which will turn minimize environmental pollution. New industries can be established which will produce value-added products from wastes [6].

Some of the wastes are identified as suitable raw materials for productin of useful ceramic materials. They are silica fumes [7], low- and high-carbon fly ash [8, 9], red mud [10], low- and high-carbon blast furnace slag [11, 12], rice-husk ash [13], Zeolites [14], Argon oxygen decarburization (AOD) slag [15], bottom ash [16], pond ash [17], man-made rock [18], biochar concrete [19], Attapulgite pozzolana [20], copper mine tailings [21], activated blast furnace slag [22], low Ca electric arc ferronickel slags [23] and metakaolin [24], etc are thought to be useful material for preparation of ceramic value-added products [6].

For preparation of useful ceramic products different methods are adopted as follows; curing [25], sol-gel [26], deep mixing [27], hydrothermal synthesis [28] and low-temperature calcination methods [29], mechanical activation (MA) [30], blended mineral admixtures [31], an alkaline solution of thermally activated silica-alumina bearing mineral with additives such as calcium carbonate [32], innovative *in situ* co-reticulation process [33], an innovative method to reclaim the waste moulding sands containing water glass with "dry" or "wet" activation of inorganic binder in waste moulding sand mixtures physically hardened by microwave radiation [34], sintering process [35], etc are adopted. Na_2CO_3, $CaCO_3$, $MgCO_3$ etc of them are used as additives.

Some related reports are as follows;

Yang Luo et al. [36] have fabricated ceramic material using clay and alkali activated pre-treated Coal fly ash (CFA). They are adopted heat-treatment process. In their work, they used low firing temperature having a wide range of sintering. Analyses of their results have shown better green strength (due to hydrogen bonding) and post-sintering performance (due to fluxing and mullite skeleton effects) if compared with ceramic tiles produced exclusively from untreated CFA.

Selvie Diana et al. [37] have fabricated Ceramic membranes composed of fly ash and clay using low sintering temperatures (900 °C and 1000°C). Membranes are prepared using different proportions of fly ash and clay i.e., 35% and 65% (M1), 45% and 55% (M2), 50% and 50% (M3), 55% and 45% (M4), 65% and 35% (M5). Pore sizes, structures, and composition are studied by SEM attached with EDX and XRD.

Shu-Hua Ma et al. [38] have reviewed fly ash, which is obtained from China's coal-fired power plants. They have studied utilization of fly ash in different areas such as extraction of valuable elements, geopolymer production, fly ash-based ceramics synthesis and soil desertification control.

Saikat Maitra et al. [39] have studied different production techniques (emerging globally) adopted to produce value added ceramics i.e., glassy materials, porcelains, refractories etc.

E. Ercenka, et al. [40] have under taken investigation on boron waste/fly ash based glass and glass-ceramics. They have used different weight percentages of boron waste powder (i.e., 10, 30, and 50 wt %) for preparation of glass and glass-ceramic materials sintering temperature of 1500 °C. Boron waste/fly ash based glass is transformed to glass-ceramic through crystallization process. Crystallization and glass-transition temperatures are examined by DTA (differential thermal analysis). Post heat treatments are done at 800 °C, 900 °C, and 1000 °C (for 1 h). The products are characterized with XRD (X-ray diffraction) analyses. Predominant phases are diopside and augite along with calconite. Hardness and fracture toughness are found out. Furthermore, Microstructure of glass-ceramic materials are examined with scanning electron microscope (SEM).

L. Carabba1 et al. [41] have developed innovative and sustainable alkali-activated composites having enhanced performance at elevated temperatures. The process developed is economically viable. For composite preparation, coal fly ash is taken as a precursor for alkali activation. Also, recycled refractory particles are used to enhance thermal stability of products. Quantitative mineralogical analyses are performed on products. Phases are identified before and after thermal exposure. Results show better dimensional stability of the composites up to 1240 °C.

Y. Yang et al. [42] have produced cost-effective porous ceramics using fly ash as main ingredient and alumina as pore-forming agent. Different parameters such as line shrink, bulk density, mechanical strength, porosity, and phase composition are determined. Properties are due to addition of alumina.

Li Zhu et al. [43] have fabricated porous mullite-whisker-structured ceramic membrane. They have used fly ash and bauxite as starting materials. Requisite amount of AlF and MoO are added to the mixture. Thereafter, membrane is prepared at a temperatures range 1100 °C to 1400 °C. AlF and/or MoO are used as catalyst and mineralizer. Porosity and pore size are also evaluated. They have examined microstructure of products and phase distribution product with the help of microscope. Porosity of membrane is measured. Mechanical performance is evaluated. Analyses of microstructure reveal elongated mullite crystals in the membrane, which is due to addition of MoO. MoO (5 wt%) has enhanced mullite formation during sintering the materials at 1200 °C. Porosity of membrane is measured to be 45.4 ± 0.9%. Similarly, 4 wt% AlF based ceramic membrane (sintered at 1200 °C) shows better open porosity (47.3 ± 0.6%) if compared with 5 wt% MoO based ceramic membrane. Mixture of MoO and AlF enhance formation of a whisker-interlocked porous structure and has improved open porosity and permeation flux.

M. Erol et al. [44] have prepared glass, glass-ceramic and ceramic materials using fly ash. They have characterized their products by X-ray diffraction (XRD) method and scanning electron microscope (SEM). Density values and toxicity characteristic are evaluated. XRD phases identified are augite phase, enstatite and mullite phases for glass, glass-ceramic, and ceramic samples. SEM images have shown tiny crystallites dispersed in the matrix of glass-ceramic. Elongated crystals are seen in ceramic samples. Density values of above material are compatible. TCLP results indicate that prepared ceramics is non-hazardous in nature. Also, ceramics prepared are inert to alkali solution.

In conclusions, microstructural, physical, chemical and mechanical properties of glass-ceramic are superior to other two ceramics (i.e., glass and ceramic samples).

Yun Fan et al. [45] have reviewed on solid by-products obtained from different sources i.e., thermal power plant, municipal solid waste, and coal.

They have synthesized zeolite by fusion-hydrothermal process using above waste materials. Process variables (i.e., NaOH/ash ratio, operating temperature and hydrothermal reaction time) are optimized.

Kanchapogu Suresh et al. [46] have fabricated ceramic microfiltration membranes using fly ash, quartz and calcium carbonate as raw material. Fabrication method employed is uni-axial dry compaction method. Raw materials and microfiltration membranes are investigated by different characterizations methods (i.e., particle size (PSD), thermogravimetric (TGA), X-ray diffraction (XRD), scanning electron microscope (SEM), mechanical stability (i.e., Young's Modulus), chemical stability, porosity, pore size and permeability). They have optimized membrane properties.

M. Senthil Kumar et al. [47] have observed abundant availability of FA, which threatening proposition to our environment. They have prepared cordierite components from fly ash (as starting material) by a low cost process. Cordierite components are extensively used for production of microelectronic components, catalyst substrate material for internal combustion engine. During synthesis of cordierite component, fly ash, magnesia and dopants (i.e., ZrO_2, CeO_2 and TiO_2) are used in different proportions (5-20 wt%). Mechanical properties (hardness, fracture toughness, and flexural strength) are determined. Thermal properties (thermal expansion coefficient, thermogravimetric and differential thermal analysis), and microstructure (scanning electron microscopy), and crystal structure (by X-ray diffraction) of prepared cordierite component are determined. Improved mechanical properties of cordierite components have made them suitable for use as catalytic substrate materials.

A. Zimmer et al. [48] have used coal fly ash from Capivari de Baixo, (Brazilian Federal State of Santa Catarina) for production of ceramic tiles. Raw materials and prepared tiles are characterized. Physico-chemical properties are evaluated.

Jian bin Zhu et al. [49] have taken coal fly ash (slurry samples) with various quantities of Al_2O_3 to fabricate mullite-based porous ceramics. They have adopted dipping-polymer-replica approach. Microstructure, phase composition, and compressive strength of prepared porous ceramics are examined. Mullite phase is identified from X-ray diffraction pattern. Both microstructure and strength properties i.e., compressive strength are strongly influenced by Al_2O_3 content.

Tomáš Húlan et al. [50] have investigated mechanical properties of tiles prepared with illitic clay, fly ash (fluidized fly ash and pulverized fly ash) in different proportions, and grog. They have followed definite pattern of heating and cooling sequence of firing. Young's moduli of prepared tiles are found to be temperature dependent. Young's moduli are obtained from dynamical thermomechanical analysis. Dimensions and mass are determined with the help of thermogravimetric and thermodilatometric analyser. In addition, flexural strength of tiles is measured using three-point bend test (at the room temperature). Results are highly encouraging.

Dong Zou et al. [51] have fabricated cost-effective ceramic membranes which have higher chemical stabilities and organic solvent resistances. Materials show promising result for water treatment applications. Cost-effective ceramic membranes from fly ash based microfiltration (MF) membrane are prepared through co-sintering process. Mullite phase have excellent heat resistance and stability and are proposed to alleviate shrinkage difference, improve porosity, and enhance bend strength. Furthermore, membrane exhibits high total organic carbon (TOC) removal efficiency (>99%) for oil-in-water (O/W) emulsion, and high stable permeability of 165 Lm h bar.

Julián Dávalos et al. [52] have produced glasses through melting of powders at 1450°C for 2 h. Further, melted powder is quenched into water to get glass-ceramic. They have evaluated temperature of glass transition temperature by differential thermal analyses (DTA). Microstructures of prepared glass-ceramics are examined. Physico-mechanical properties and durability (acidic and alkaline environments) are also evaluated. Results are highly encouraging and therefore glass-ceramics is a good candidate for construction applications.

Yelong Zhao et al. [53] have produced ceramic tiles, bricks and blocks by using fly ash. They have investigated effect of alkali on preparation process of foam material. They have evaluated water absorption, apparent density and compressive strength of their prepared products. Experimental results have displayed homogenous microstructures with interconnected pores, good water absorption behaviour, good apparent density, and high compressive strength values.

Haihui Mi et al. [54] have prepared ultra-light ceramic foams by green spheres technique, using waste glass powder and fly ash. During the preparation of glass materials, borax and SiC are used as fluxing agent and foaming agent. Process parameters such as Fly ash, borax, and sintering temperature have effects on microstructures, properties of ceramic foams. Ultra-light ceramic foams are formed between 680 to 780°C. It owns bulk density (0.14–0.41 g/cm^3), porosity (82.9–94.1%), compressive strength (0.91–6.37 MPa), and thermal conductivity (0.070–0.121 W m K). It is concluded that green sphere method is convenient, cost-effective, and environment friendly.

Gayatri Sharma et al. [55] have reviewed on the production of fly ash. They are compared it with volcanic ash and natural soil. They conclude that fly ash is suitable for use in ceramic industries.

Shuming Wang et al. [56] have produced unique high performance glass-ceramics from fly ash. They have descried methodologies of products. They have also identified principal crystal phases, and have correlated properties with possible application of these materials one discussed in details.

2.2 Characterization Techniques

Usually, characterizations such as XRD, FTIR, UV Visible, SEM/EDS, and electrical analysis (two probe electrical resistivity), etc are needed to investigate the evidence and required performance of the prepared ceramics from thermal power plant wastes.

2.2.1 X-Ray Diffraction (XRD) Technique

X-ray diffraction (XRD) is an important non-destructive technique, which is used to govern structure [57], of as prepared materials It is also assessed different structural parameters such as crystal planes (hkl), unit cell angle, unit cell dimension (a, b, c), inter layered-spacing (d), etc. X-rays are produced from target CuKα having wavelength (λ) and X-ray energy is obtained from the relation [58, 59]

$$E = h\upsilon = h\frac{c}{\lambda} \qquad (2.1)$$

Where,
  h is the Planck's constant (6.62×10^{-34} joule)
  c is the velocity of light (3×10^{-8} m/s)
  E is the energy of the radiation.
  Wavelength of X-ray is comparable to the size of atoms.

X-ray diffraction is mainly based on constructive interference of monochromatic X-rays and prepared samples. The interaction between incident X-rays and sample produces constructive interference. Diffracted rays are produced and satisfies the Bragg's law (nλ=2dsinθ). By scanning the sample in the a range of 2θ angles, all possible diffraction directions of the lattice should be attained due to the random orientation of the powdered materials. Conversion of the diffraction peaks to d-spacing allows the identification of the materials because each material has a set of unique d-spacing. Typically, this is achieved by comparison of d-spacing with standard reference patterns.

In the current thesis work, X-ray diffraction patterns are recorded using CuKα radiation (wavelength, λ=0.154 nm) with fix the operation parameters i.e., 40 kV and 20 mA. Powder samples are put on a quartz sample holder (at room temperature) and are scanned at diffraction angle 2θ from 10° to 70°.

2.2.2 Fourier Transformation Infra-Red (FTIR) Spectroscopy

FTIR is also another non-destructive technique (NDT) is required to characterize the prepared materials for identifying presence of chemical groups. The technique is used to analyse the presence of functional groups, formation of chemical linkage between used materials and other phases with have been investigated using a FTIR spectrometer (Nexus-870, Thermo Nicolet Corp, and USA). FTIR instrument parameters are fixed (50 scan at 4 cm^{-1} resolution, transmittance/absorbance mode). The spectrometer [60, 61] is used IR radiations, which is obtained from an IR source and are passed through the sample. Amount of IR energy adsorbed/transmitted is recorded by suitable detector and is guided through an interferometer where a Fourier Transform is performed on the output signal.

In this characterization, as prepared powder samples are used for this measurement, the powdered samples are prepared by making pellet using both small amounts of as prepared samples and KBr. Primarily; KBr granules are grinded with mortar pestle. After that, small amount of samples are placed and further grinded the mixture. Then, finally grinded samples is put in die and then placed in the hydraulic pressure. It is compacted the samples by pressure (5 kg.f). Pellet (13 mm diameter, 0.3 mm thick) so prepared is prepared and used for IR characterization. Background spectrum is collected before running the samples. Prepared pellets samples are put in a sample holder. Pellets are exposed to IR radiation in the spectrometer and data are collected. This technique is completed to characterize the bonding type of the molecules and for each type of bonding it produced characteristic absorption bands.

2.2.3 Scanning Electron Microscopy (SEM)

Surface morphologies of prepared materials are analysed by SEM. SEM is a microscope that uses electrons in place of light to produce surface

CERAMIC PRODUCTION METHODS AND CHARACTERIZATION TECHNIQUES

image [62]. In the analysis, electron beam produced from electron gun and is focused on a small portion of the sample. Vacuum is maintained. Detector collects the output signals during the interaction of electrons with the sample. The signal is sent to a computer and forms the final image. Thus, particular preparation technique is required for the sample preparation before their analysis. All nonconducting materials need thin layer of gold or platinum coating. Coating is completed by 'sputter coater'. During the coating, operating voltage is kept constant (i.e. 4 kV). Such coater uses an electric field and argon gas. Sample is placed in the coating chamber. Argon gas is ionized in the applied electric field to form argon ion (Ar^+). The argon ions knock gold atoms from the surface of the gold foil and get deposited on sample. Particularly,

The SEM is also capable of performing semi-quantitative chemical composition analysis by energy dispersive X-ray spectroscopy (EDXS) and wavelength dispersive X-ray spectroscopy (WDXS) analysis [62]. Data are collected over a selected area of the surface of the sample, and a 2-dimensional image is generated which displays spatial variations in these properties. Areas ranging from approximately 1 cm to 5 microns in width can be imaged in a scanning mode using conventional SEM techniques (magnifications ranging from 20x to approximately 30,000x with spatial resolution of 50 to 100 nm).

2.2.4 Electrical Characterizations

2.2.4.1 Electrical Resistivity Measurement by Two Probe (at Room Temperature)

Resistivity measurement at room temperature as well as temperature dependent (presence and absence of magnetic field) is an important electrical characterization to the present work.

Resistivity value is found, which signify the electrical nature of materials. For this measurement a two probe set-up is used as shown in Figure 2.1.

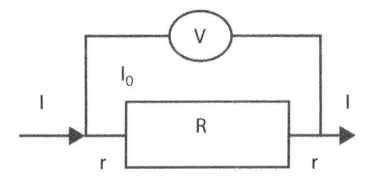

Figure 2.1 Schematic representation of linear two probe set-up.

Usually, resistivity of semiconductor is measured by two probes method. It is a one of the standard electrical measurement method. In this investigation, linear two probe technique is used to measure both room and low temperature resistivity. Pellet samples are taken for this work. Two electrical contacts are made by attaching copper wires onto the sample surface through silver paste. In this technique, two probes contact points are arranged linearly in a straight line at equal distance (S) from each other. For room temperature measurement, the sample is fixed on insulating plate where four probes are connected. The output voltage is related proportionally to the applied current. Such type of contacts was called ohmic. According to two point probe method, the resistivity (ρ) was calculated using the relation [63].

$$\rho = \frac{RA}{l} \quad (2.2)$$

Where S is the probe spacing in centimetre (cm), which was kept constant, I is the supplied current in millampere (mA) or nanoampere (nA) and the corresponding voltage was measured in volt (V) or millivolt (mV). The conductivity (σ) is calculated using the relation [63].

$$\sigma = \frac{1}{\rho} \quad (2.3)$$

For low temperature resistivity measurement, sample is placed in a specific chamber and the pressure is maintained at 10^{-5} torrs. The Lakeshore (model 331) temperature controller is connected. Then, DC resistivity is measured. A constant current is passed through the one probe and measured voltage in the other probes. A DC current source (Keithley 220 programmable) is taken. Different millampere/nanoampere current is applied and corresponding voltage was noted.

2.3 Conclusions

General background information on ceramic is discussed in this chapter. Literatures relevant to thermal power plant based ceramics are presented in brief. Various raw materials and different production methods of ceramic are reviewed in detail. It is emphasized that there lies a major gap in research which deals with fruitful utilization of industrial waste. Therefore, a research goal is fixed to develop new ceramic product using thermal power plant waste.

Acknowledgements

First author would like to thank Prof. Munesh Chandra Adhikary, PG Council Chairman, Fokir Mohan University, for his invaluable guidance, advice, and constant inspiration throughout the entire program. First author is willing to thank Mr Mukteswar Mohapatra in Fokir Mohan University for his support. Authors convey their sincere thanks to GIET, University Gunupur, Rayagada, Odisha, India for providing Lab facilities to do the research work. Authors also like to thank the CRF, IIT Kharagpur for providing their testing facilities.

References

1. (a). Heimann, Robert B, (2010), Classic and Advanced Ceramics: From Fundamentals to Applications, https://web.archive.org/web/20201210175959/https://books.google.com/books?id=JiJsfP_DiL4C&q=ceramics+are&pg=PR15.
(b). C. B. Carter, and M. G. Norton, (2007). Ceramic Materials: Science and Engineering. Springer, pp. 20 & 21, ISBN 978-0-387-46271-4.
(c). keramiko/s . Liddell, Henry George; Scott, Robert; A Greek–English Lexicon at the Perseus Project https://www.perseus.tufts.edu/hopper/text?doc=Perseus:text:1999.04.0057:entry=ker amiko/s)5.ke/ramos, https://www.perseus.tufts.edu/hopper/text?doc=Perseus:text:1999.04.0057:entry=ke/r.
2. H.-P. Martin, (2014), Conductive Ceramics as Electrical Materials at High Temperatures, pp. 38, Annual Report 2014/15.
3. Ceramics and Glass in Everyday Life, the American Ceramic Society. https://ceramics.org/about/what-are-engineered-ceramics-and-glass/ceramics-and-glass in everyday-life, Accessed on 2019 Jun 30.
4. R. William, (1978), Properties of ceramic raw materials, 2nd Ed. Guildford. Surrey: Pergamon international library; 1978.
5. W. E. Worrall, (1986), Clays and Ceramic Raw Materials. 2nd Ed. London: Elsevier applied science publishers; 1986.
6. N. Abduganiev, (2020), the Use of Thermal Technologies for the Recovery of Value Added Products from Household Solid Waste: A Brief Review, *IOP Conf. Series: Earth and Environmental Science*, Vol. 614, 012005.
7. E. M. M. Ewais, Y. M. Z. Ahmed, and A. M. M. Ameen, (2009), Preparation of Porous Cordierite Ceramic using A Silica Secondary Resource (Silica

Fumes) for Dust Filtration Purposes, *Journal of Ceramic Processing Research*, Vol. 10, pp. 721-728.

8. T. Sathanandam, P. O. Awoyera, V. Vijayan, and K. Sathishkumar, (2017), Low carbon building: Experimental insight on the use of fly ash and glass fibre for making geopolymer concrete, *Sustainable Environment Research*, Vol. 27, pp. 146-153 https://doi.org/10.1016/j.serj.2017.03.005.

9. B. Cetin, A. H. Aydilek, and Y. Guney, (2010), Stabilization of Recycled base Materials with High Carbon Fly Ash, *Resources Conservation and Recycling*, Vol. 54, pp. 878-892.

10. H. Sun, C. Chen, L. Ling, S. Ali Memon, Z. Ding, W. Li, L.-P. Tang, and F. Xing, (2019), Synthesis and Properties of Red Mud-Based Nanoferrite Clinker, *Journal of Nanomaterials*, Vol. 2019, pp. 1-12 https://doi.org/10.1155/2019/3617050.

11. N. You, J. Shi, and Y. Zhang, (2020), Corrosion Behaviour of Low-Carbon Steel Reinforcement in Alkali-Activated Slag-Steel Slag and Portland Cement-Based Mortars Under Simulated Marine Environment, *Corrosion Science*, Vol. 175, pp. 108874-xxx https://doi.org/10.1016/j.corsci.2020.108874.

12. I. Narasimha Murthy, N. Arun Babu, and J. Babu Rao, (2016), High Carbon Ferro Chrome Slag – Alternative Mould Material for Foundry Industry, *Procedia Environmental Sciences*, Vol. 35, pp. 597-609 https://doi.org/10.1016/j.proenv.2016.07.046.

13. L. Sun and K. Gong, (2001), Silicon-Based Materials from Rice Husks and Their Applications, *Industrial & Engineering Chemistry Research*, Vol. 40, pp. 5861–5877 https://doi.org/10.1021/ie010284b.

14. M. Granda-Valdés, A. I. Pérez-Cordoves, and M. E. Díaz-García, (2006), Zeolites and Zeolite-based Materials in Analytical Chemistry, *TrAC Trends in Analytical Chemistry*, Vol. 25, pp. 24-30 https://doi.org/10.1016/j.trac.2005.04.016.

15. E.-J. Moon and Y. C. Choi, (2019), Carbon Dioxide Fixation via Accelerated Carbonation of Cement-Based Materials: Potential for Construction Materials Applications, *Construction and Building Materials*, Vol. 199, pp. 676-687 https://doi.org/10.1016/j.conbuildmat.2018.12.078.

16. M. Cabrera, J. L. Díaz-López, F. Agrela, and J. Rosales, (2020), Eco-Efficient Cement-Based Materials using Biomass Bottom Ash: A Review, *Applied Science*, Vol. 10, pp. 8026-24.

17. M. K. Panigrahi, P. K. Rana, A. K. Pradhan, P. K. Rout, A. K. Samal, S. Gupta and Mv B. Kumar, (2015), Production of Geopolymer based Construction Material from Pond Ash: An Industrial Waste, *19th International Conference Non-Ferrous Metal-2015* (9, 10th July), pp. 190-200.

18. Davidovits, Joseph. (1994). Geopolymers: Man-made rock geosynthesis and the resulting development of very early high strength cement. *J. Mater. Edu.*, 16. 91-137.

19. L. Chen, Y. Zhang, L. Wang, S. Ruan, J. Chen, H. Li, J. Yang, V. Mechtcherine, and D. C. W. Tsang, (2022), Biochar-augmented carbon-negative concrete,

Chemical Engineering Journal, Vol. 431, pp. 133946 https://doi.org/10.1016/j.cej.2021.133946.

20. T. Shi, Y. Liu, Y. Zhang, Y. Lan, Q. Zhao, Y. Zhao, and H. Wang, (2022), Calcined Attapulgite Clay as Supplementary Cementing Material: Thermal Treatment, Hydration Activity and Mechanical Properties, *International Journal of Concrete Structures and Materials*, Vol. 16, pp. 1-10 https://doi.org/10.1186/s40069-022-00499-8.

21. R. S. Krishna, F. Shaikh, J. Mishra, G. Lazorenko, and A. Kasprzhitskii, (2021), Mine tailings-based geopolymers: Properties, applications and industrial prospects, *Ceramics International*, Vol. 47, pp. 17826-17843 https://doi.org/10.1016/j.ceramint.2021.03.180.

22. T. Luukkonen, Z. Abdollahnejad, J. Yliniemi, P. Kinnunen, and M. Illikainen, (2018), Comparison of Alkali and Silica Sources in One-Part Alkali-Activated Blast Furnace Slag Mortar, *Journal of Cleaner Production*, Vol. 187, pp. 171-179 https://doi.org/10.1016/j.jclepro.2018.03.202.

23. K. Komnitsas, D. Zaharaki, and V. Perdikatsis, (2007), Geopolymerization of Low Calcium Ferronickel Slags, *Journal of Materials Science*, Vol. 42, pp. 3073-3082.

24. P. G. Asteris, P. B. Lourenço, P. C. Roussis, C. E. Adami, D. J. Armaghani, C. L. Cavaleri, E. Chalioris, M. M. Hajihassani, E. Lemonis, Ahmed S. Mohammed and K. Pilakoutas, (2022), Revealing the Nature of Metakaolin-Based Concrete Materials using Artificial Intelligence Techniques, *Construction and Building Materials*, Vol. 322, pp. 126500-xxx https://doi.org/10.1016/j.conbuildmat.2022.126500.

25. Z. Chen, D. Li, W. Zhou, and L Wang, (2010), Curing Characteristics of Ceramic Stereolithography for an Aqueous-Based Silica Suspension, *Proceedings of the Institution of Mechanical Engineers Part B Journal of Engineering Manufacture*, Vol. 224, pp. 641-651.

26. A. Nazeri, E. Bescher, and John D. Mackenzie, (1993), Ceramic Composites by the Sol-Gel Method: A Review, *Ceramic Engineering & Science Proceedings*, Vol. 14, pp. 1-19.

27. Sk S. Hossain and P.K. Roy, (2020), Sustainable Ceramics Derived from Solid Wastes: A Review, *Journal of Asian Ceramic Societies*, Vol. 8, pp. 984-1009 https://doi.org/10.1080/21870764.2020.1815348

28. J. Ortiz, C. Gomez-Yañez, R. López-Juárez, I. A. Velasco, and D. H. Pfeiffer, (2012), Synthesis of Advanced Ceramics by Hydrothermal Crystallization and Modified Related Methods, *Journal of Advanced Ceramics*, Vol. 1, pp. 204-220.

29. P. Perumal, A. Hasnain, T. Luukkonen, P. Kinnunen, and M. Illikainen, (2021), Role of Surfactants on the Synthesis of Impure Kaolin-Based Alkali-Activated, Low-Temperature Porous Ceramics, *Open Ceramics*, Vol. 6, pp. 100097-9 https://doi.org/10.1016/j.oceram.2021.100097.

30. V. Berbenni, A. Marini, and G. Bruni, (2001), Effect of Mechanical Activation on the Preparation of $SrTiO_3$ and Sr_2TiO_4 Ceramics from the Solid State Systems $SrCO_3$-TiO_2, *Journal of Alloys and Compounds*, Vol. 329, pp. 230-238.
31. M. Amin, B. A.Tayeh and I. S. Agwa, (2020), Effect of using Mineral Admixtures and Ceramic Wastes as Coarse Aggregates on Properties of Ultrahigh-Performance Concrete, *Journal of Cleaner Production*, Vol. 273, 10 pp. 123073-xxx https://doi.org/10.1016/j.jclepro.2020.123073.
32. T. Jiang, G. Li, G. Qiu, X. Fan, and Z. Huang, (2008), Thermal Activation and Alkali Dissolution of Silicon from Illite, *Applied Clay Science*, Vol. 40, pp. 81-89.
33. F. Colangelo, G. Roviello, L. Ricciotti, C. Ferone and R. Cioffi, (2013), Preparation and Characterization of New Geopolymer-Epoxy Resin Hybrid Mortars, Materials, Vol. 6, pp. 2989-3006.
34. R. R. Mishra and A. K. Sharma, (2016), Review on Microwave-Material Interaction Phenomena: Heating Mechanisms, Challenges and Opportunities in Material Processing, *Composites Part A: Applied Science and Manufacturing*, Vol. 81, pp.78-97 https://doi.org/10.1016/j.compositesa.2015.10.035.
35. S. V. Fomichev, N. P. Dergacheva, A. V. Steblevskii, and V. A. Krenev, (2011), Production of Ceramic Materials by the Sintering of Ground Basalt, *Theoretical Foundations of Chemical Engineering*, Vol. 45, pp. 526-529.
36. Y. Luo, S. Zheng, S. Ma, C. Liu, and X. Wang, (2017), Ceramic Tiles Derived from Coal Fly Ash: Preparation and Mechanical Characterization, Ceramics International, Vol. 43, pp.11953-11966.
37. F. Razi and M. R. Bilad, S. Diana, R. Fauzan, and N. Arahman, (2020), Synthesis And Characterization of Ceramic Membrane from Fly Ash and Clay Prepared by Sintering Method at Low Temperature, *Rasayan Journal of Chemistry*, Vol. 13, pp. 1335-1341 http://dx.doi.org/10.31788/RJC.2020.1335707.
38. S.-H. Ma, M.-D. Xu, Qiqige, X.-H. Wang, and X. Zhou, (2017), Challenges and Developments in the Utilization of Fly Ash in China, *International Journal of Environmental Science and Development*, Vol. 8, pp. 781-785.
39. S. Maitra, (1999), Ceramic Products from Fly Ash Global Perspectives, Central Fuel Research Institute, Dhanbad, Fly Asia Utilisation for Value Added Products Eds. B. Chatterjee, K. K. Singh & N. G. Goswami, 1999, NML, Jamshedpur, pp. 32-37.
40. E. Ercenka, U. Sena, G. Bayrakb and S. Yilmaza, (2014), Glass and Glass-Ceramics Produced from Fly Ash and Boron Waste, *Acta Physica Polonica A*, Vol. 125, pp. 626-628.
41. L. Carabba, S. Manzi, E. Rambaldi, G. Ridolfi, M. C. Bignozzi, (2017), High-Temperature Behaviour of Alkali-Activated Composites based on Fly Ash and Recycled Refractory Particles, *Journal of Ceramic Science and Technology*, Vol. 8, pp. 377-388.

42. Y. Yang, F. Liu, Q. Chang, Z. Hu, Q. Wang and Y. Wang, (2019), Preparation of Fly Ash-Based Porous Ceramic with Alumina as the Pore-Forming Agent, *Ceramics*, Vol. 2, pp. 286-295 https://doi.org/10.3390/ceramics2020023.
43. L. Zhu, Y. Dong, L. Li, J. Liu and S.-J. You, (2015), Coal Fly Ash Industrial Waste Recycling For Fabrication Of Mullite-Whisker-Structured Porous Ceramic Membrane Supports, *RSC Advances*, Vol. 5, pp. 11163-11174 (https://doi.org/10.1039/C4RA10912K).
44. M. Erol, S. Küçükbayrak, and A. Ersoy-Meriçboyu, (2008), Comparison of the Properties of Glass, Glass-Ceramicand Ceramic Materials Produced from Coal Fly Ash, *Journal of Hazardous Materials*, Vol. 153, pp. 418-25.
45. Y. Fan, F.-S. Zhang, J. Zhu, and Z. Liu, (2008), Effective Utilization of Waste Ash from Msw and Coalco-Combustion Power Plant: Zeolite Synthesis, *Journal of Hazardous Materials*, Vol. 153, pp. 382-8.
46. S. Kanchapogu, G. Pugazhenthi, and R. Uppaluri, (2016), Fly Ash based Ceramic Microfiltration Membranes for Oil-Water Emulsion Treatment: Parametric Optimization using Response Surface Methodology, *Journal of Water Process Engineering*, Vol. 13, pp. 27-43 https://doi.org/10.1016/j.jwpe.2016.07.008.
47. M. S. Kumar, M. Vanmathi, G. Senguttuvan, R. V. Mangalaraja and G. Sakthivel, (2019), Fly Ash Constituent-Silica and AluminaRole in the Synthesis and Characterization of Cordierite Based Ceramics, *Silicon*, Vol. 11, pp. 2599-2611 https://link.springer.com/article/10.1007/ s12633-018-0049-0.
48. A. Zimmer and C. P. Bergmann, (2007), Fly ash of Mineral Coal as Ceramic Tiles Raw Material, Waste Management, Vol. 27, pp. 59-68.
49. J.-b. Zhu and H. Yan, (2017), Microstructure and Properties of Mullite-Based Porous Ceramics Produced from Coal Fly Ash with Added Al_2O_3, *International Journal of Minerals, Metallurgy, and Materials*, Vol. 24, pp. 309-315 https://doi.org/10.1007/s12613-017-1409-2.
50. T. Húlan, I. Štubna, J. Ondruška and A. Trník, (2020), The Influence of Fly Ash on Mechanical Properties of Clay-Based Ceramics, *Minerals*, Vol. 10, pp. 930-12.
51. D. Zou, M. Qiu, X. Chen, E. Drioli, and Y. Fan, (2019), One Step Co-Sintering Process for Low-Cost Fly Ash Based Ceramic Microfi Ltration Membrane in Oil-in-Water Emulsion Treatment, Separation and Purification Technology, Vol. 210, pp. 511-520 https://doi.org/10.1016/j.seppur.2018.08.040.
52. J. Dávalos, A. Bonilla, M. A. Villaquirán-Caicedo, R. M. de Gutiérrez, and J. Ma. Rincónb, (2020), Preparation of Glass-Ceramic Materials from Coal Ash and Rice Husk Ash: Microstructural, Physical and Mechanical Properties, Boletín de la Sociedad Española de Cerámica y Vidrio, Vol. xxx, pp.1-11.
53. Y. Zhao, J. Ye, X. Lu, M. Liu, Y. Lin, W. Gong, and G. Ning, (2010), Preparation of Sintered Foam Materials By Alkali-Activated Coal Fly Ash, *Journal of Hazardous Materials*, Vol. 15, pp.174(1-3):108-12.

54. H. Mi, J. Yang, Z. Su, T. Wang, Z. Li, W. Huo, and Y. Qu, (2017), Preparation of Ultra-Light Ceramic Foams from Waste Glass and Fly Ash, *Advances in Applied Ceramics (Structural, Functional and Bioceramics)*, Vol. 116, pp. 400-408.
55. G. Sharma, S. K. Mehla, T. Bhatnagar, and A. Bajaj, (2013), Possible Use of Fly Ash in Ceramic Industries: An Innovative Method to Reduce Environmental Pollution, International Conference on Ceramics, Bikaner, India, *International Journal of Modern Physics: Conference Series*, Vol. 22, 99-102.
56. S. Wang, C. Zhang, and J. Chen, (2015), Utilization of Coal Fly Ash for the Production of Glass-ceramics With Unique Performances: A Brief Review, *Journal of Materials Science & Technology*, Vol. 30, pp. 1208-1212 https://www.jmst.org/article/2014/1005-0302-30-12-1208.htm.
57. M., Alexandre, and P. Dubois, (2000), Polymer-layered Silicate Nanocomposites: Preparation, Properties and Uses of A New Class of Materials, *Materials Science and Engineering R*, Vol. 28, pp. 1-63.
58. H. P. Klugg, and L. E. Alexander, (1974), X-ray Diffraction Procedures; John Wiley & Sons: U.S.A.
59. D. Moore, and Jr. R. Reynolds, (1997), X-ray Diffraction and Identification and Analysis of Clay Minerals of Clay Minerals; Second Ed.; Oxford University Press, U. K.
60. FTIR Spectroscopy Attenuated Total Reflectance (ATR), www.perkinelmer.com.
61. Smith, W. E., and Dent, G. (2005), Modern Raman Spectroscopy-A Practical Approach, John Wiley & Sons, Ltd, ISBNs: 0-471-49668-5 (HB); 0-471-49794-0 (PB).
62. Carl Zeiss Microscopy, (2004), Detection Principles based on GEMINI® Technology.
63. P. P. Sengupta, P. Kar, and B. Adhikari, (2009), Influence of Dopant in the Synthesis, Characteristics and Ammonia Sensing Behavior of Processable Polyaniline, *Thin Solid Films*, Vol. 517, pp. 3770-3775.

3

High Resistance Sintered Fly Ash (FA) Ceramics

M. K. Panigrahi[1]*, R.I. Ganguly[2] and R.R. Dash[3]

[1]*Department of Materials Science, Maharaja Sriram Chandra BhanjaDeo University, Balasore, Odisha, India*
[2]*Department of Metallurgical Engineering, National Institute of Technology, Raurkela, Odisha, India*
[3]*CSIR-National Metallurgical Laboratory, Jamshedpur, Jharkhand, India*

Abstract

One of the industrial waste materials i.e., Fly Ash (FA) has been used as a source material for development of a novel product. Product will help to use for different applications, particularly, in electrical sectors.

Samples are made by mixing Fly ash with dextrin (as binder) and sintered at different temperatures (900 °C, 1100 °C, and 1300 °C). Effects of sintering temperature on formation of different phases are studied by X-ray Diffraction method. With increasing sintering temperature, Mullite phase formation is favored due to interaction between different constituents of FA. Mullite phase promotes electrical insulating property and hence, there is a strong possibility for development of insulator from FA. SEM photographs show fused FA particles. Inter-particular spaces between fused mass are much reduced and thereby, increase density of mass. FTIR analyses of sintered FA products indicate Si-O-Al band, Si-O band, and H-O-H band, respectively. With increasing sintering temperature, resistance increases from 28.5 Mega ohms to 35.1 Mega ohms. It is inferred from present work that good insulating material can be produced from FA.

Keywords: Fly ash, mullite, insulator, XRD, FESEM, resistivity

---

Corresponding author: muktikanta2@gmail.com

3.1 Introduction

Conversion of raw materials such as fly ash (FA) into a value added product has become an important challenge for R & D organizations. FA pollutes environment as well as spoil cultivable land [1–5]. Therefore, use of this material is thought to be of prime interest for R & D organizations. Conversion of this material into value added products will be advantageous [6–8].

Usually, FA is generated from thermal power plants. All these problems have led researchers to come forward to utilize FA for developing tiles [9], refractories [10], insulators [11], cutting tools [12], etc.

Since constituents of FA are mainly Silica (SiO_2) and Alumina (Al_2O_3), therefore, there is scope for developing an insulator which is useful for electrical industries. Utilization of ceramic materials such as electrical insulators goes back to 1850. Werner von Siemens [13] has used ceramics during the construction of electrical air lines. There are numerous characteristic behaviours of ceramics *i.e.*, mechanical strength, high dielectric strength, and good corrosion resistance. Electrical insulators can be classified on the basis of their purpose and insulation properties [14, 15]. In Today's scenario, there is growing demand for low cost insulators in the area of electrical engineering fields. This is due to increasing demand because of higher consumption of electric energy in modern society. Therefore, researchers are motivated to develop cost-effective insulator which will serve the purpose of different technological applications [16].

Usually refractories are used for the following purposes; furnace lining [10], cutting tools [12], insulator [11], etc. Generally, these are prepared through compaction followed by firing between 1000 °C-1400 °C. It is possible to produce a Mullite phase (Al_2SiO_5). Mullite phase enhances resistance of compacted products [17].

Prospective uses of ceramics in the field of electrical engineering are felt necessary. Since Dielectric materials developed from it will help to produce capacitors and transducers helpful for electrical applications.

Ceramic capacitors are being used as electronic components. In the present day, fast growing capacitor technology is progressing with ceramic based materials. Products will meet for microelectronics and communication engineering. It is expected that ceramic capacitors will be used more in the near future [18].

From an electrical properties point of view, Insulators have great importance for day-to-day life. It makes lives easier and shock free. Electrical pin, a kind of insulator, is used for electrical poles [6–8]. This materials are also required in high tech areas such as defence [19, 20], capacitors [21], supercapacitors [22], microwave engineering [23], phase shifter [24], filter [25] and resonator [26] etc.

In the present work, Fly ash is sintered at temperatures 900 °C, 1100 °C, 1200 °C, and 1300 °C. Morphological studies with EDS analysis are done. Different phases such as quartz, Cristobalite, TbO_2, and mullite of sintered Fly ash materials are identified by X-ray diffraction analyses. Fourier transform infrared spectroscopy (FTIR) analysis is carried out to recognize different chemical groups in sintered materials. Finally, electrical resistivity values of sintered materials have been estimated.

3.2 Experimental Details

3.2.1 Materials and Chemicals

Fly ash is received from NALCO, Damanjodi, Odisha. Composition of Fly ash is indicated in Table 3.1. Dextrin $(C_6H_{10}O_5)_n \cdot xH_2O$ is procured from Loba chemicals. 6-10% Water is added to prepare samples.

Table 3.1 Different constituents with percentage of as-received NALCO fly ash [27].

	Used industrial waste
	Fly ash
Constituents with percentages	Fe_2O_3 = 8.1 MgO =1.14 Al_2O_3 =24.98 SiO_2 =55.85 P_2O_5 =0.15 SO_3 =1.16 K_2O =0.85 CaO =2.54 Na_2O =0.2 TiO_2 =1.75 CO_2=1.56
Sources	NALCO, BBSR
References	[27]

3.2.2 Materials Preparation

Fly ash is grinded in a motar with pastel for 2 h and is sieved through 240 meshes. Then, the fly ash is dried in heating oven at 120 °C for 2 h to remove moisture.

10 g sieved Fly ash (FA) is further grinded in a motar with pastel for 2 h. Dextrin (0.05%) is used as binder to bind FA particles. Again, the mixture is grinded for 1 h. Small amount of water is added to make mixture pasty. Pellets are made from paste. Pellet is termed as green sample.

For preparation of pellets, pressure (10 MPa) is applied by universal testing machine (UTM) to the pelletizer. After that, pelletizer is removed from the system and then, green pellets are removed. Pellet is now ready for sintering.

For sintering, pellets are placed in a muffle furnace and bisquetted at 900 °C for 2 h. During bisquetting, water and dextrin are removed from

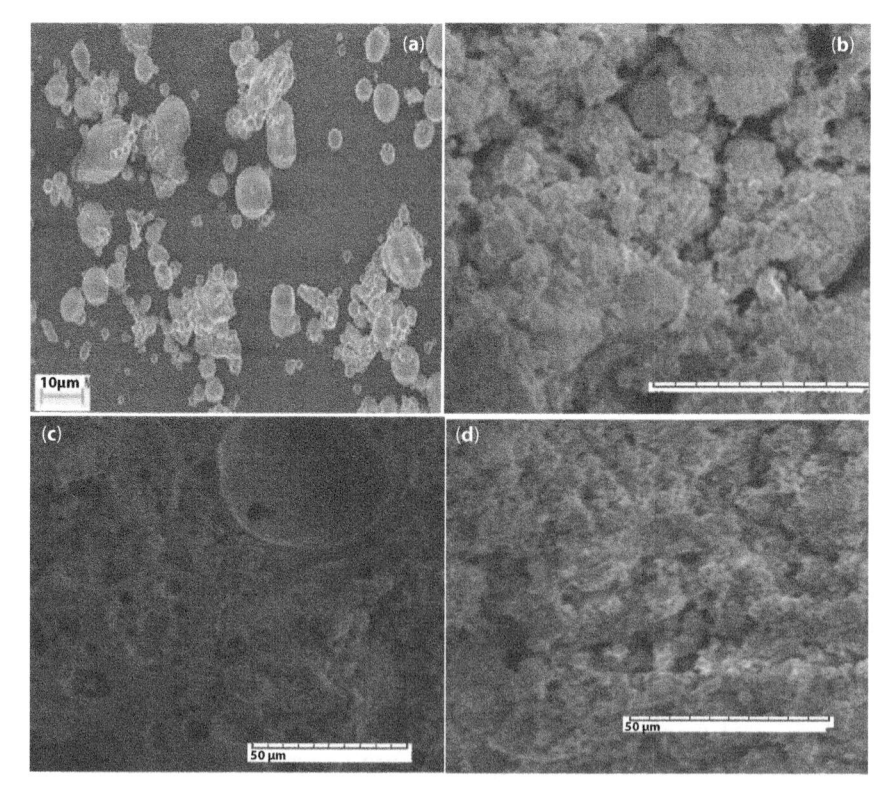

Figure 3.1 Morphological investigation of as-received fly ash (a), FA sintered at 900 °C (b), FA sintered at 1100 °C (c), FA sintered at 1200 °C (d) [27].

Table 3.2 Different set of sintered fly ash based materials preparation.

S. no.	Individual composition name		
	Fly ash (%)	Dextrin (%)	Sintering temperatures (°C)
1	100	0.05	900
2	100	0.05	1100
3	100	0.05	1200

pellets. Then, pellets are sintered at 1200 °C for 1 h. Entire process is described in Figure 3.1.

After sintering, pellets are allowed to be cooled to room temperature slowly in the furnace and then, they are removed. Average sintered pellet dimensions (*i.e*, thickness and diameter) are 0.5 cm and 3 cm, respectively [27].

Similar process is carried out at sintered temperature 1000 °C and 1100 °C (Vide-Table 3.2). Sintered pellet(s) are ready for characterizations.

3.2.3 Physical Characterizations

X-ray diffraction analyses are made for samples to identify different phases. For this work, Phillips PW-1710 advanced wide angle X-ray diffractometer, Phillips PW-1729 X-ray. Samples are loaded on a quartz sample holder and are placed in diffractometer. Samples are scanned at diffraction angle (2θ) ranging bctwccn 20° to 180° with a scanning spccd 2°/min.

Surface topography of sintered samples is analyzed by field emission scanning electron microscope (Carl Zeiss Supra 40). For scanning analyses, samples are coated with gold using sputtering technique. FESEM operating voltage is kept at 30 kV.

FTIR spectra of sintered FA are recorded on a Thermo Nicolt Nexus 870 spectrophotometer. The analyses are made from 400 cm^{-1} to 4000 cm^{-1}. Measurements are taken in absorbance mode. Instrument parameters are kept constant (50 scan at 4cm^{-1} resolution) during the measurements. For these analyses, samples (*i.e.*, pellet) are prepared by compression molding using a pelletizer. Potassium Bromide (KBr) and small amounts of test samples are used to make pellets. A background spectrum is collected before doing experiments. Then, a pellet is put in a sample holder and data are collected.

Electrical measurements are made for sintered pellets. Resistivity is estimated using expression [27].

$$\rho = \frac{R \times A}{l} \qquad (3.1)$$

Here, **A** is the area of electrode $= \frac{\pi}{4} D^2$ (m^2), **R** is the resistance of prepared samples (MΩ), **l** is thickness of the sample, **ρ** is the resistivity of the samples in MΩ.m.

3.2.4 Results and Discussion

FESEM images of fly ash and sintered Fly ash are shown in Figures 3.1(a, b, c, d). Surface topography of Fly ash is shown Figure 3.1(a). Fly ash shows irregular elliptical shaped particles. Figure 3.1(a) describes particles scattered in the matrix with a few spots congregation of particles wide spread in the matrix. Particles are of varied sizes. Dimension of large particles have major axis (~13 micron) and minor axis (~6-7 micron). The smallest particle shows dimension having a major axis (~1 micron) and a minor axis (~0.5 micron). In general, aspect ratio varied between ~1.8 to 2.0.

There are marked differences observed in these topological diagrams (Figures 3.1a, 3.1b, and 3.1c). In general, sintered samples show fusion of particles having some pores in it. As a result of sintering, FA particles are fused together with reduction of average interparticle spacing. As sintering temperature increases from 900 °C to 1200 °C, the product has become denser in nature with lesser voids present in the microstructure.

EDS analyses are done for three sintered samples along with FESEM studies (Figures 3.2a, 3.2b, and 3.2c). In Figure 3.2a, Elements such as O, Al, Si, Ti, Fe, and K, are present in Fly Ash except Na, Ca, and Mg. Ti is present in trace amounts in sintered fly ash.

O, Al, Si, Ti, Fe, K, Na, Ca, and Mg are found in sintered Fly Ash (at 1200 °C, Figure 3.2c). Ti and Mg elements are present in trace amounts in sintered Fly Ash (at 1200 °C, Figure 3.2c). All the elements (except Mg) are shown in Figure 3.2b.

Figure 3.3 shows XRD pattern of Fly Ash. Identified different phases are alumina, silica, and mullite. The highest obtained intensity phase is silica. Other two predominant phases are mullite and alumina (Figure 3.3) [27].

Figures 3.4a, 3.4b, and 3.4c are XRD diffraction pattern of FA sintered at 900 °C, 1100 °C, and 1200 °C, respectively. Details of predominant phases are shown in Table 3.3. Table 3.3 specifies crystal system, space group, space group number, cell dimension, cell angle, density of unit cell, respectively.

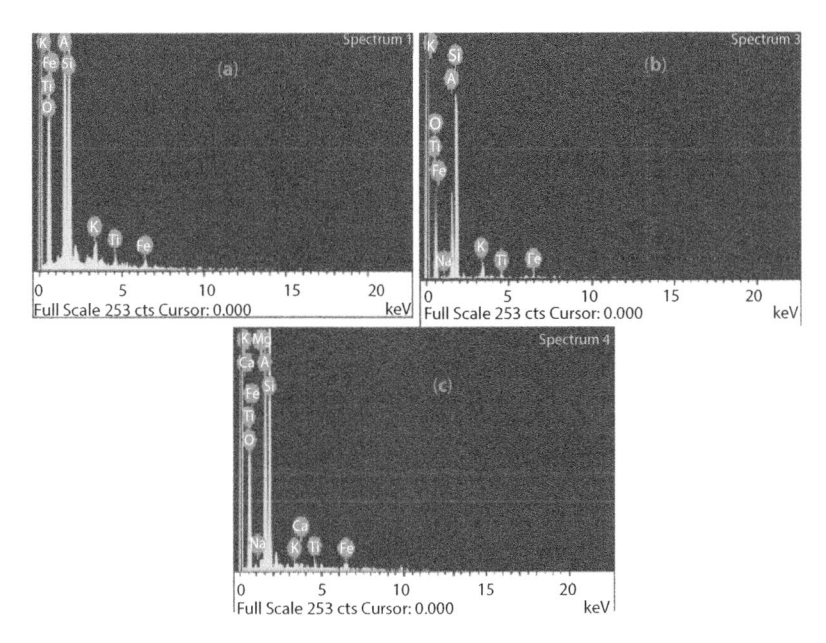

Figure 3.2 EDS spectrum of fly ash sintered at 900 °C (a), 1100 °C (b), 1200 °C (c) [27].

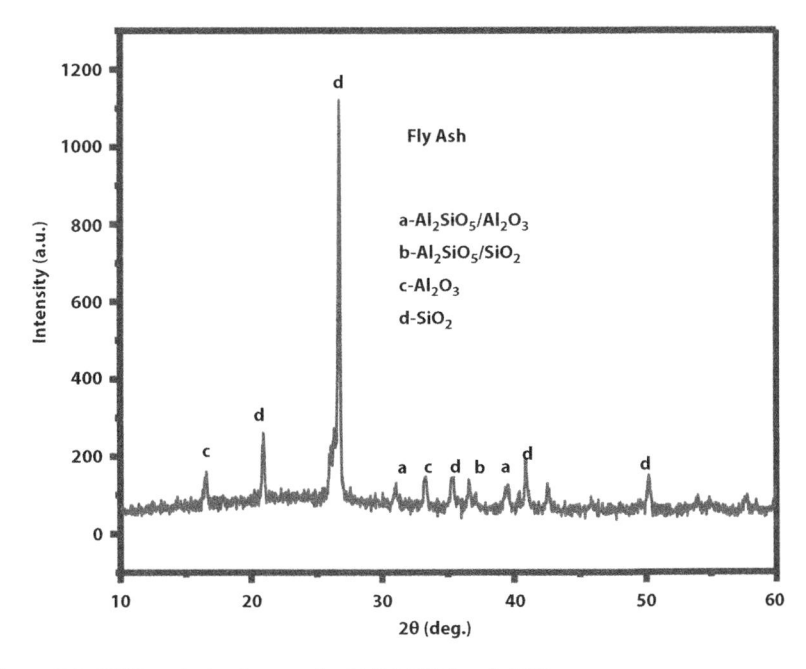

Figure 3.3 XRD analysis of as-received NALCO fly ash [27].

a-Al_2SiO_5/Al_2O_3
b-Al_2SiO_5/SiO_2
c-Al_2O_3
d-SiO_2

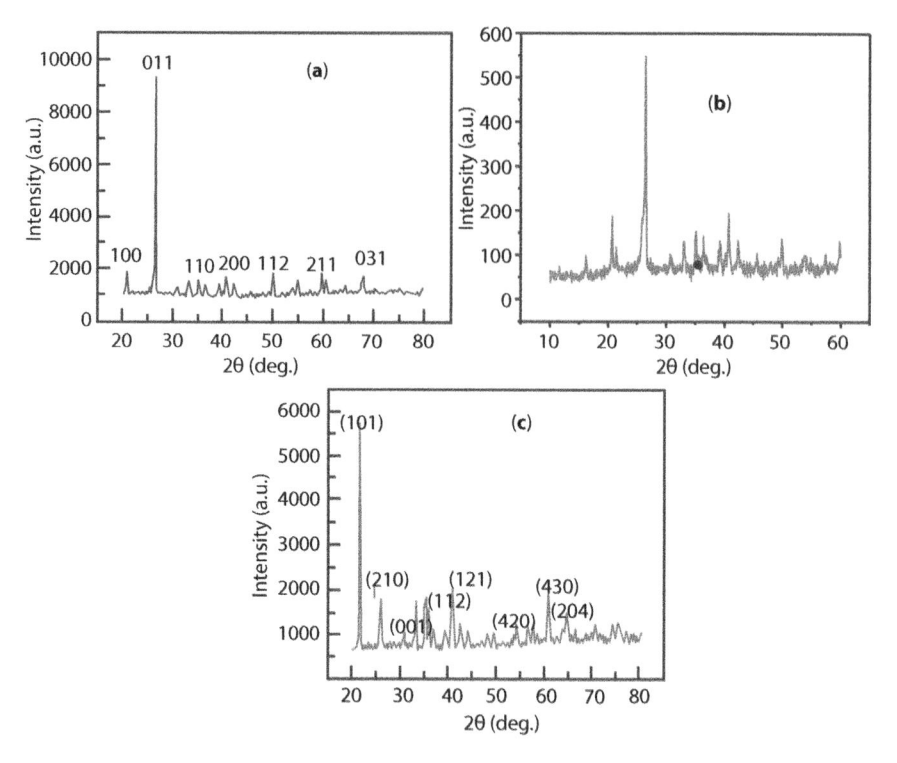

Figure 3.4 (a) XRD pattern of fly ash sintered at 900 °C (Quartz Phase) [27]. (b) XRD pattern of fly ash sintered at 1100 °C [27]. (c) XRD pattern of fly ash sintered at 1200 °C [27].

It can be observed that all the sintered samples have mullite phase. However, it is observed that amount of mullite phase increase with increase in sintering temperature. Alumina and silica binary phase diagram suggests that mullite phase dominate at 1300 °C.

FTIR absorption bands of as-processed NALCO fly ash is studied [22–24]. Presence of different absorption stretching frequency of Fly-ash powder corresponds to Fly-ash components. Stretching frequency at 600 cm^{-1} matches the Si-O-Al band [28]. Stretching frequency at 1098 and 1608 cm^{-1} are accredited to Si-O-Si asymmetric band [29] and H-O-H, respectively.

Superimposed FTIR spectrum of the Fly ash based sintered materials (sintering temperature 1000 °C, 1100 °C, and 1200 °C) are shown in Figure 3.5. Different peaks of the spectrum show presence of different absorption bands of sintered material. Figure 3.5 shows occurrence of various type absorption bands arising at different wave numbers *i.e.*, 3441, 2918, 1823, 1638, 1117 and 907 cm^{-1}. Stretching bands at 3441 cm^{-1} and 1638 cm^{-1} indicate vibrations of O-H stretching bands and H-O-H bending bands,

Table 3.3 Crystallographic parameters of sintered fly ash [27].

Crystal ID	Mineral name	Crystal system	Space group	Space group number	Cell dimension	Crystal angle	Calculated density (g/cm³)
FA_900 °C	Quartz	Hexagonal	P3221	154	a=b≠c	α=β=90°, γ=120°	15.89
FA_1100 °C	Mullite	Orthorhombic	Pbam	55	a≠b≠c	α=β=γ=90°	3.17
FA_1100 °C	Terbium Oxide	Cubic	I213	199	a=b=c	α=β=γ=90°	7.89
FA_1200 °C	Mullite	Orthorhombic	Pbam	55	a≠b≠c	α=β=γ=90°	3.16
FA_1200 °C	Cristobalite	Tetragonal	P41212	92	a=b≠c	α=β=γ=90°	2.26

Note: FA-Fly Ash.

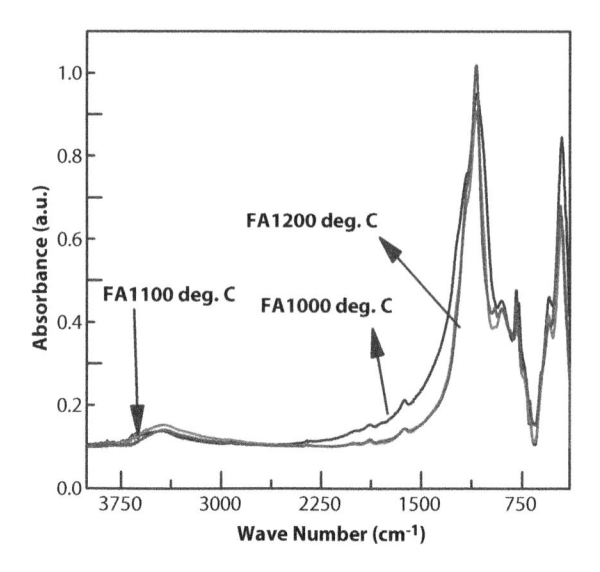

Figure 3.5 FTIR spectrum of sintered fly ash at different temperatures (1000 °C and 1200 °C) [27].

respectively [29]. Absorption peaks at 1117 cm^{-1} shown by Si-O band and signifies presence of silicate groups. Al^{3+}O^{2-} absorption bands are also seen at 907 cm^{-1} [30].

Resistivity value of as-received fly ash and sintered Fly Ash is mentioned in Table 3.4. From resistivity data, Fly ash sintered at 1200 °C shows the highest resistivity and is measured to be 496.12095 × 10^6 Ω.cm,

Table 3.4 Resistivity value of as-received fly ash and processed fly ash [27] [radius=1.5 cm and thickness=0.5 cm].

S. no.	Sample ID	Processing temperature (°C)	Resistance value (MΩ)	Resistivity value (Ω.cm)
1	Fly Ash [28]	As received	28.5	402.83325 × 10^6
2	Processed Fly Ash	900	28.7	405.66015 × 10^6
3	Processed Fly Ash	1100	30.7	433.92915 × 10^6
4	Processed Fly Ash	1200	35.1	496.12095 × 10^6

whereas as-received Fly ash shows lowest resistivity and is determined to be 402.83325×10^6 Ω.cm. This is due to formation of mullite in different quantities at different sintering temperatures [28].

3.3 Conclusions

Fly ash is sintered at four different temperatures i.e., 900 °C, 1000 °C, 1100 °C, and 1200 °C. Microstructure of fly ash shows irregular spherical spheres of oxide particles dispersed in the matrix. Due to sintering, there are distinct microstructural changes if compared with the microstructure of unsintered fly ash. XRD patterns have identified two distinct phases i.e., mullite and quartz. Intensity pattern of XRD shows formation if higher amount of mullite with increasing sintering temperature. FTIR measurement suggests three distinct stretching frequencies Si-O-Al, Si-O, and H-O-H. The corresponding groups of the peaks indicate formation of above bonds within the phases. Maximum electrical resistivity (Fly ash sintered at 1200 °C) is found to be 35.1 MΩ for material sintered at 1200 °C, which is due to formation of higher amount of mullite.

Acknowledgements

First author would like to thank Prof. Munesh Chandra Adhikary, PG Council Chairman, Fokir Mohan University, for their invaluable guidance, advice, and constant inspiration throughout the entire program. First author would also like thank Mr. Mukteswar Mohapatra in Fokir Mohan University for his support. Authors convey their sincere thanks to GIET, University Gunupur, Rayagada, Odisha, India for providing Lab facilities to do the research work. Authors would also like to thank the CRF, IIT Kharagpur for providing their testing facilities.

References

1. N. S. Pandian, C. Rajasekhar, and A. Sridharan, (1998), Studies of the Specific Gravity of Some Indian Coal Ashes, *Journal of Testing and Evaluation*, Vol. 26, 177.
2. N. S. Pandian, (2004), Fly Ash Characterization with Reference to Geotechnical Applications, *Journal of the Indian Institute of Science*, Vol. 84, 189.

3. M. Ahmaruzzaman, (2010), A review on the Utilization of Fly Ash, *Progress in Energy and Combustion Science*, Vol. 36, 327.
4. L. Wang, G. Hu, F. Lyu, T. Yue, H. Tang, H. Han, Y. Yang, R. Liu, W. Sun, (2019), Application of Red Mud in Wastewater treatment, *Minerals*, Vol. 9, 281.
5. S. Sushil and V.S. Batra, (2008), Catalytic Applications of Red Mud, an Aluminium Industry Waste: A Review, *Applied Catalysis B: Environmental*, Vol. 81, 64.
6. C. L. Goldsmith, A. Malczcwski, J. J. Yao, S. Chen, J. Ehmk, and D. H. Hinzel, (1999), RFMEMS-Based Tunable Filters, *International Journal of RF and Microwave Computer-Aided Engineering*, Vol. 9, 362.
7. G. Subramanyam, F. V. Keuls and F. A. Miranda, (1998), Novel K-band Tunable Microstrip v-band Pass Filter using Thin Film HTS/Ferroelectric/Dielectric Multilayer Configuration, *IEEE Microwave Guided Wave Letter*, Vol. 8, 78.
8. A. Tombak, J.P. Maria, F.T. Ayguavives, Z. Jin, G.T. Stauf, A.I. Kingon, A. Mortazawi, (2003), Voltage-Controlled RF Filters Employing Thin Film Barium-Strontium Titanate Tunable Capacitors, *IEEE Transaction Microwave Theory and Technology*, Vol. 51, 462.
9. A. Olgun, Y. Erdogan, Y. Ayhan, and B. Zeybek, (2005), Development of Ceramic Tiles from Coal Fly Ash and Tincal Ore Waste, *Ceramics International*, Vol. 31, pp.153-158.
10. D. Goski and M. Lambert, (2019), Engineering resilience with precast monolithic refractory articles, *International Journal of Ceramic Engineering and Science*, Vol. 1, pp. 169-177.
11. M. Zhu, R. Ji, Z. Li, H. Wang, L. L. Liu, and Z. Zhang, (2016), Preparation of Glass Ceramic Foams for Thermal Insulation Applications from Coal Fly Ash and Waste Glass, *Construction and Building Materials*, Vol. 112, pp. 398-405.
12. L. Li and Y. Li, (2017), Development and trend of ceramic cutting tools from the perspective of mechanical processing, *IOP Conference Series: Earth and Environmental Science*, Vol. 94 pp. 012062-6.
13. W. V. Siemens, (1966), Inventor and Entrepreneur: Recollections of Werner Von Siemens, London, England.
14. T. Tanaka and T. Imai, (2013), Advances in Nanodielectric Materials over the past 50 years, *IEEE Electrical Insulation Magazine*, Vol. 1, pp. 10-23.
15. T. Tanaka, (2005), Dielectric Nanocomposites with Insulating Properties, *IEEE Transactions on Dielectrics and Electrical Insulation*, Vol. 12, pp. 914-928.
16. E. C. Nzenwa and A. D. Adebayo, (2019), Analysis of Insulators for Distribution and Transmission Networks, *American Journal of Engineering Research (AJER)*, Vol.8, pp. 138-145.
17. K. Cui, Y. Zhang, T. Fu, J. Wang, and X. Zhang, (2020), Toughening Mechanism of Mullite Matrix Composites: A Review, *Coatings*, Vol. 10, pp. 672-696.

18. R. Kulke, G. Mollenbeck, C. Gunner, P. Uhlig, K. H. Drue, S. Humbla, J. Muller, R. Stephan, D. Stopel, J. F. Trabert, G. Vogt, M. A. Hein, A. Molke, T. Baras, A. F. Jacob, D. Schwanke, J. Pohlner, A. Schwarz, and G. Reppe, (2009), Ceramic Microwave Circuits for Satellite Communication, *Journal of Microelectronics and Electronic Packaging*, Vol. 6, pp. 27-31.
19. M. M. Harussani, S. M. Sapuan, G. Nadeem, T. Rafin, and W. Kirubaanand, (2022), Recent applications of carbon-based composites in defence industry: A review, *Defence Technology*, Vol. 18, pp. 1281-1300.
20. https://www.syalons.com/2019/01/08/ceramic-materials-in-defence-applications/
21. L. Y. V. Chen, R. Forse, D. Chase, and R. A. York, (2004), Analog Tunable Matching Network using Integrated Thin-Film BST Capacitors, *IEEE Microwave Theory and Technology Society*, Vol. 1, 261.
22. M. I. A. Abdel Maksoud, R. A. Fahim, A. E. Shalan, M. Abd Elkodous, S. O. Olojede, A. I. Osman, C. Farrell, A. H. Al-Muhtaseb, A. S. Awed, A. H. Ashour, and David W. Rooney, (2021), Advanced Materials and Technologies for Supercapacitors Used in Energy Conversion and Storage: A Review, *Environmental Chemistry Letters*, Vol. 19, pp. 375-439.
23. J. Nath, D. Ghosh, J.P. Maria, A.I. Kingon, W. Fathelbab, D. Paul, F.Z. Michael, B. Steer, (2005), An Electronically Tunable Microstrip Band Pass Filter using Thin-Film Barium-Strontium Titanate (BST). *IEEE Trans. Microwave Theory and Technology*, Vol. 53, 2707.
24. M. Pu, L. Liu, W. Xue, Y. Ding, H. Ou, K. Yvind, and J. M. Hvam, (2010), Widely Tunable Microwave Phase Shifter based on Silicon on Insulator Dual-Microring Resonator, Optics Express, Vol. 18, pp. 6172-6182, https://doi.org/10.1364/OE.18.006172.
25. X.-L. Lü and H. Xie, (2019), Spin Filters and Switchers in Topological-Insulator Junctions, *Physical Review Applied*, Vol. 12, pp. 064040-10.
26. F.H. Wee1, F. Malek, S. Sreekantan, A.U. Al-Amani, F. Ghani1, and K.Y. You, (2011), Investigation of the Characteristics of Barium Strontium Titanate (BST) Dielectric Resonator Ceramic Loaded on Array Antennas, *Progress in Electromagnetic Research*, Vol. 121, 181.
27. M.K. Panigrahi, (2022), Investigation of Structures Of Sintered Fly Ash Materials: Resources of Industrial Wastes, *Bulletin of Scientific Research*, Vol. 4, pp. 1-10.
28. T. Tanaka, (2005), Dielectric Nanocomposites with Insulating Properties, *IEEE Transactions on Dielectrics and Electrical Insulation*, Vol. 12, pp. 914-928.
29. B. J. Saikia, G. K Rao, and P. Sarathy, (2010), Fourier Transform Infrared Spectroscopic Characterization of Kaolinite from Assam and Meghalaya, North Eastern India, *Journal of Morden Physics*, Vol. 1, 206.
30. R. P. dos Santos, J. Martins, C. Gadelha, B. Cavada, A. V. Albertini, F. Arruda, M. Vasconcelos, E. Teixeira, F. Alves, J. L. Filho, and V. Freire, (2014), Coal Fly Ash Ceramics: Preparation, Characterization, and Use in the Hydrolysis of Sucrose, *Scientific World Journal*, Vol. 2014, 01.

4
High Resistance Sintered Fly Ash/Kaolin (FA/CC) Ceramics

M. K. Panigrahi[1*], R.I. Ganguly[2] and R.R. Dash[3]

[1]*Department of Materials Science, Maharaja Sriram Chandra Bhanja Deo University, Balasore, Odisha, India*
[2]*Department of Metallurgical Engineering, National Institute of Technology, Raurkela, Odisha, India*
[3]*CSIR-National Metallurgical Laboratory, Jamshedpur, Jharkhand, India*

Abstract

Paper describes electrical properties of ceramic insulator prepared with fly ash (FA) and kaolin (CC). Kaolin is mixed with Fly ash in different proportions i.e., 10, 20, 30, 40, and 50 %. For the preparation of composite, uniaxial pressure (i.e., 10 MPa) has been applied to samples followed by sintering of mixture at 1200 °C. It is found that 40 % china clays mixed with fly ash has yielded maximum resistivity i.e., 39.5×10^7 Ωm. This composite is believed to be highly competive in insulator market. Prepared composite has been characterized with x-ray diffraction, field emission scanning electron microscope with energy dispersive analyser, Fourier transformation infrared spectroscope, dielctric, and thermo gravimetric analysis. Dielectric value of the composite is estimated at room temperature in the frequency range of 1-500 kHz. The resistivity and dielectric properties of developed composite has indicated enormous potential of the material for electrical applications as insulator.

Keywords: Industrial wastes, Fly Ash (FA), Kaolin (CC), Mullite, Quartz, resistivity, dielectric

---

Corresponding author: muktikanta2@gmail.com

4.1 Introduction

Chemical and physical properties of Fly ash suggest that they can be utilized to develop some value-added products. By utilization of these environmental hazardous materials, human health and soil are protected [1–5]. Hence, scientific communities are making considerable efforts to use industrial wastes for making useful materials. Compositions of this material suggest production of electrical materials from this waste [6–8]. From application point of view, Insulators have great importance for everyday life and they provide safety to our devices. Electrical pin used in electrical poles is also a kind of insulator [6–8]. Similarly dielectrics are very important for instruments, which are required for day to day life, high-tech applications i.e., defence [9], microwave [10], capacitors [11]. Insulators/dielectrics have been developed as a tunable material which has a great application in microwave engineering [10]. So, they can be used as a phase shifter [12], filter [13] and resonator [14] etc. Further, they can also be used in manufacturing of radomes. Such types of materials are in-valuable for defence applications [9], and manufacturing electronic instruments. Thus, development of these materials (insulators/dielectrics) from industrial waste is thought to be good proposition for producing electronic appliances.

In the present work, fly ash is mixed with kaolin in different proportions and bisquetted at 900 C for 2 h. Then, it is sintered at 1200 °C. After sintering materials are characterized by XRD, FESEM attached to EDS analyser. Electrical properties i.e., resistivity and Frequency dependence dielectric constant (k) are determined at room temperature. Samples are also investigated through thermogravimetric analyser.

4.2 Experimental Section

4.2.1 Materials and Method

Fly ash is collected from NALCO, Damanjodi, Odisha. Kaolin is purchased from Merck, India. Dextrin $(C_6H_{10}O_5)_n.xH_2O)$ is purchased from Loba chemicals. Water (6%) is added to prepare green sample(s). The chemical composition of Fly ash and kaolin are presented in Table 4.1.

Table 4.1 Chemical composition of fly ash and kaolin [15].

	Used industrial waste	Clays
	Fly ash	Kaolin
Constituents with percentages	Fe_2O_3 = 8.1 MgO =1.14 Al_2O_3 =24.98 SiO_2 =55.85 P_2O_5 =0.15 SO_3 =1.16 K_2O =0.85 CaO =2.54 Na_2O =0.2 TiO_2 =1.75 CO_2=1.56	Fe_2O_3= 1.54 Al_2O_3 =37.66 SiO_2 =44.80 MgO =Trace K_2O =NA CaO =0.50 Na_2O =Trace TiO_2 =0.60 ZrO_2=Trace L.O.I=14.33
Sources	NALCO, BBSR	Merck, India

4.3 Preparation of Test Samples

4.3.1 Preparation of Sintered FA/CC Composite

Fly ash is grinded in a ball mill for 5 h and is sieved through 240 meshes. Then, the fly ash is dried in a heating oven at 120 °C for 2 h to remove moisture. Different sets of FA/CC composite are prepared (vide-Table 4.2). For each set, appropriate proportion of Fly ash and kaolin are mixed together and are grinded. 0.05 % dextrin and 6-10% water is added to the mixture before grinding. Grinding is continued for 1 h, which ensures formation of homogenous mixture. Mixture so prepared is called green mixture. Mixture is now ready for making pellets.

Pellets are prepared by universal testing machine (UTM) with pelletizer. Assembled pelletizer is cleaned before use. 10 MPa pressure is maintained to pelletizer for five minutes. After compaction, pellet sample is removed from the pelletizer. Thicknesses and diameter of as-prepared pellets are measured to be 0.5 cm and 3 cm, respectively. Green pellets are bisquetted at 900 °C for 1 h followed by sintering at 1200 °C for 2 h. During bisquetting, dextrin and intacted water are removed. Entire process of composite preparation is shown by chart, which is described in Figure 4.1. These sintered samples are ready to be used for different tests [15].

Table 4.2 Compositions (%) of fly ash/CC composite materials preparation (sintered at 1200 °C).

S. no.	Individual composition name	
	Fly ash	Kaolin (CC)
1	50	50
2	60	40
3	70	30
4	80	20
5	90	10

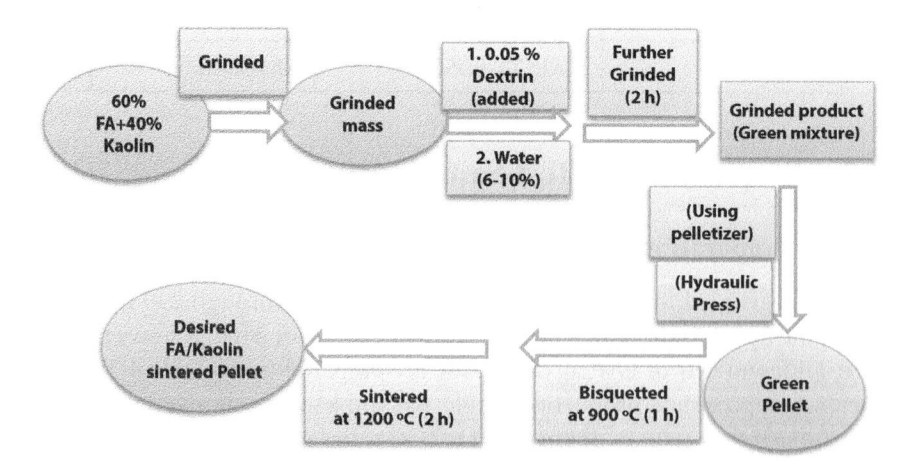

Figure 4.1 Sintered fly ash (60%)/kaolin (40%) composite preparation flow chart [15].

4.4 Characterization Techniques

X-ray diffraction experiments of FA, CC, and sintered FA/CC composite are carried out in Phillips PW-1710 advance wide angle X-ray diffractometer. Powder samples are placed on a quartz sample holder and are scanned through diffraction angles (2θ) varying between 10° to 70° with scanning speed 2°/min.

FTIR spectra of FA, CC, and sintered FA/CC composite are recorded on a Thermo Nicolt Nexus 870 spectrophotometer in the range 400-4000 cm^{-1}. FTIR test sample (i.e., pellets) is prepared by compression molding method using a pelletizer. Small amounts of test powder samples are mixed with potassium Bromide (KBr). Before taking the test, background spectrum is recorded.

Surface morphologies of FA, CC, and sintered FA/CC composite are analysed by field emission scanning electron microscope (FESEM, Carl Zeiss Supra 40). Before FESEM experiment, samples are gold coated via sputtering technique.

Dielectric values of sintered FA/CC composite are measured at room temperature using impedance analyzer (PSM1735 model, Newtons4th Ltd, UK). For this measurement, pellet samples (circular surface) are taken. Probe is contacted on aluminium foils. The diameter of probe is 2.8 mm. Dielectric constant (k) is calculated using the relation [15].

$$k = \frac{C \times d}{A\varepsilon_0} \quad (4.1)$$

Where 'A' is area of electrode $= \frac{\pi}{4}D^2$ (m^2), 'd' is thickness of prepared samples (m), 'ε_0' is permittivity in vacuum = 8.854 × 10^{-12} F/m and C is capacitance in farad. Frequency range is 10 Hz to 500 kHz.

TGA analyses of sintered FA/CC composite are performed in Perkin Elmer Pyris Diamond analyzer. The test is done in nitrogen environment. Heating rate is 10 °C/min.

4.5 Results and Discussion

Figure 4.2 shows three XRD pattern of Kaolin (a), fly ash (b), and Sintered Fly ash/Kaolin composite (c). Phases identified for CC (Figure 4.2a) shows presence keoline and quartz. Keoline is a dominating phase marked by higher peaks. Thus, Kaolin contains kaolins and quartz. Phases identified of Fly ash are quartzite and mullite. Quartzite is a predominant phase in comparison to mullite phase. XRD pattern of Sintered mixture (60FA-40CC) shows number of peaks. This indicates that there is transformation of more crystalline phases from amorphous phases present in fly ash before sintering of the mixture. This is corroborated with FESEM microstructure. Dominated phase in sintered materials are quartzite and mullite phases. These observations are interesting because mullite is a well-known electrical and thermal insulator, which is used in many industrial and commercial purposes.

FTIR spectra of the Fly ash and FA/CC composite material are shown in Figure 4.3. FTIR spectra indicate presence of different absorption bands in each category of material. Figure 4.3 indicates presence of different

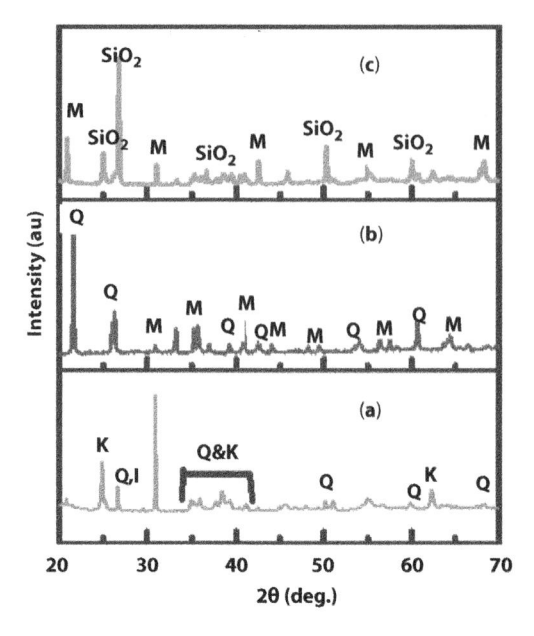

Figure 4.2 XRD plot of kaolin (a), fly ash (b), sintered fly ash/kaolin composite (c) [16] [*Note: M-Mullite, Q-Quartz, K-Kaolinite, I-Illite] [15].

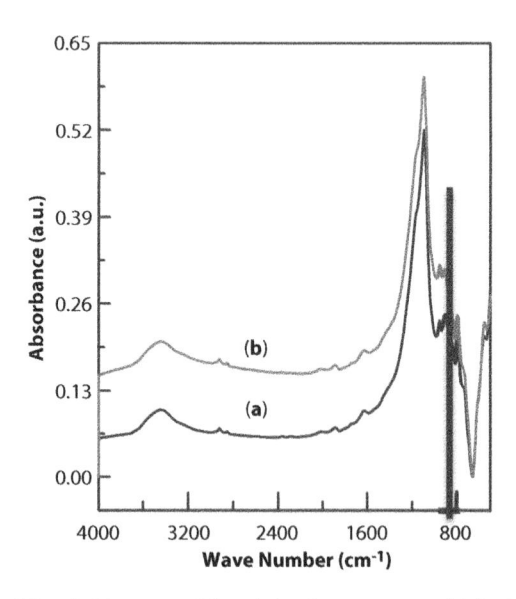

Figure 4.3 FTIR of fly ash (a), sintered fly ash/kaolin composite (b) [15].

absorption bands occurring at different wave numbers *i.e.*, 3441, 2918, 1823, 1638, 1117 and 907 cm^{-1}. Bands are in agreement with stretching vibrations of O-H bonds (3441 cm^{-1} wave number) and H-O-H bending vibrations (1638 cm^{-1} wave number) of interlayer adsorbed H$_2$O molecule [16]. Hydroxyl-stretching band of water plays an important role and peak shift of the FTIR spectra is significant. Absorption band at 1117 cm^{-1} wave number is attributed to Si-O band and signifies the occurrence of silicate groups. Presences of Al^{3+}O^{2-} absorption bands are also indicated near 907 cm^{-1} wave number [17].

FESEM images of Fly ash (A), Kaolin (B), and sintered Fly ash/Kaolin composite (at 1200 °C) are shown in Figure 4.4(a-c). It can be seen from

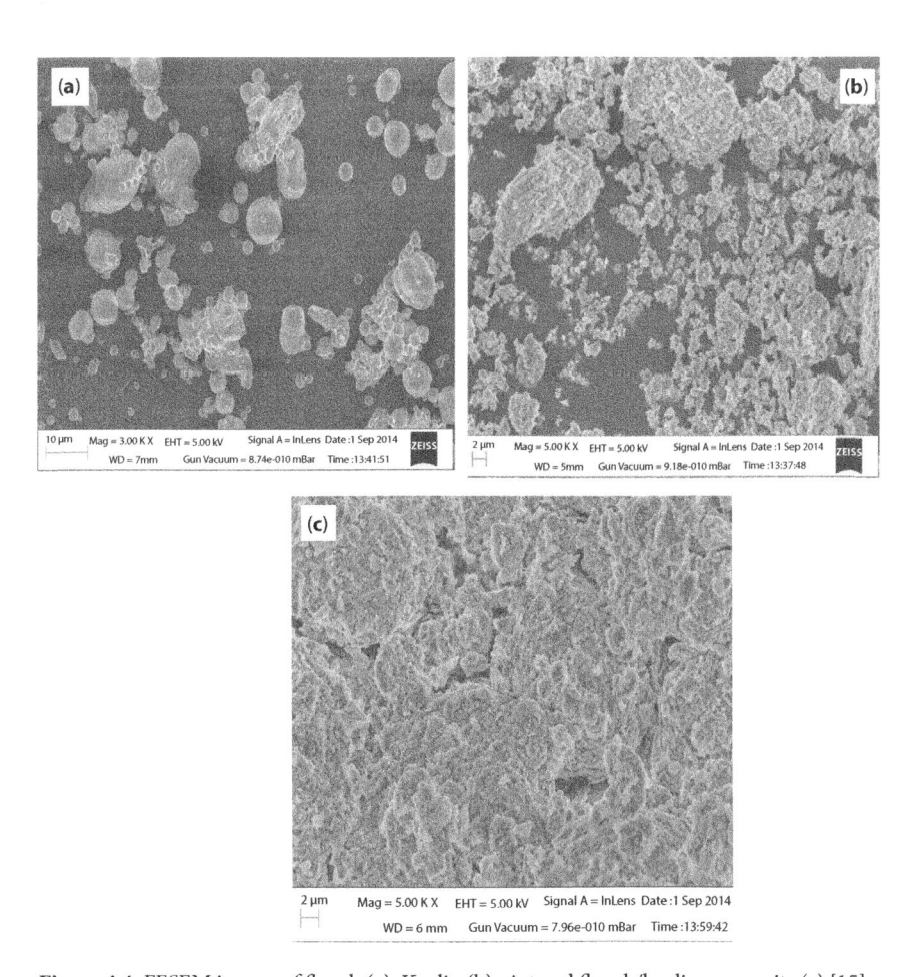

Figure 4.4 FESEM images of fly ash (a), Kaolin (b), sintered fly ash/kaolin composite (c) [15].

FESEM image of FA (Figure 4.4a) that Fly Ash mainly constitutes by irregular shaped of spheres with a smoother texture. Also, some quartz particles, residue of un-burnt coal or some vitreous unshaped fragments can be seen. In Figure 4.4b, it shows FESEM micrograph of kaolin, which is flaky in nature. FESEM image of sintered Fly ash/kaolin composite shows non-uniform segregated structure with small pores. This is formed due to interaction of kaolin (CC) and spheres of FA particles. It is expected to be advantageous due to higher crystallinity.

EDS analyses are done during the FESEM studies (Figure 4.5). Elements such as O, Al, Si, Fe, Ca, and Mg are present both in kaolin, Fly Ash, and the composite except Na, Ti,, and K elements (Figure 4.5). Na, Ti, and K elements are present in trace amounts in fly ash and kaolin. Occurrence of all those elements in the composite is due to reaction kaolin with fly ash during sintering process.

Table 4.3 shows resistivity value for fly ash, kaolin, and composites prepared with different proportions of FA and CC. Both FA and CC show

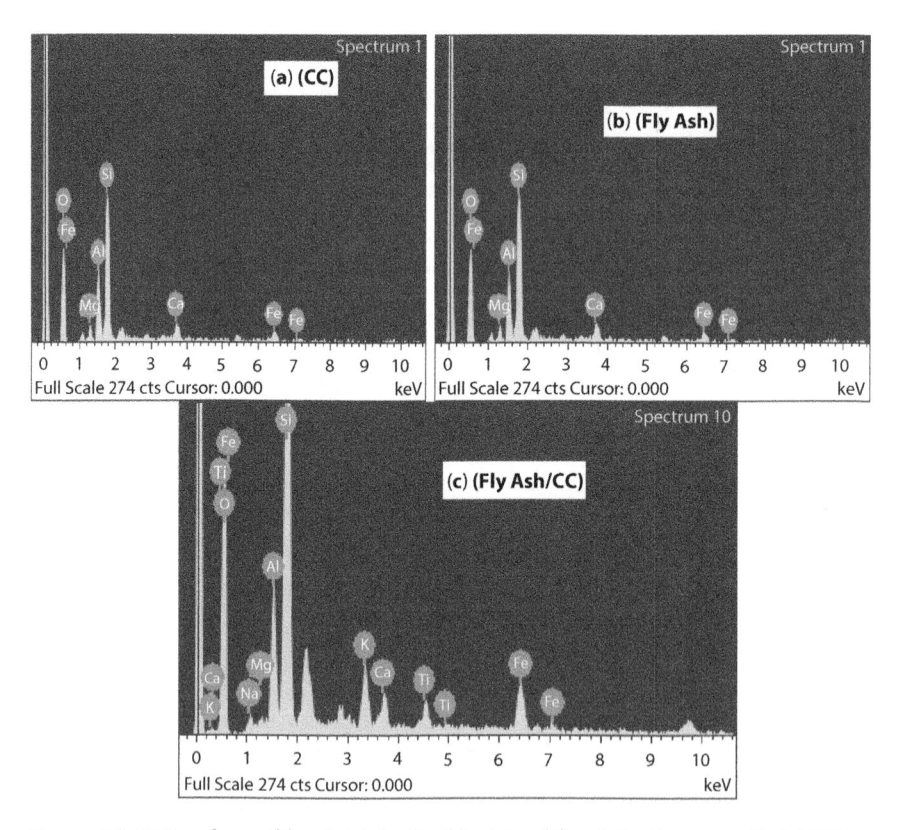

Figure 4.5 EDS analyses of fly ash (a), kaolin (b), sintered fly ash/kaolin composite (c) [15].

Table 4.3 Resistivity value of fly ash, kaolin, fly ash/kaolin composites prepared with different compositions [15].

S. no.	Sample ID	Resistivity (Ωm)
1	S0 (FA100:CC0) [20]	2.85×10^7
2	S1 (FA0:CC100) [21]	0.73×10^7
3	S2 (FA90:CC10) at 1200 °C	7.36×10^7
4	S2 (FA80:CC20) at 1200 °C	12.1×10^7
5	S3 (FA70:CC30) at 1200 °C	17.5×10^7
6	S4 (FA60:CC40) at 1200 °C	39.5×10^7
7	S5 (FA50:CC50) at 1200 °C	27.03×10^7

resistivity 2.85×10^7 and 0.73×10^7 Ωm), respectively. Prepared composites show many fold higher resistivity if compared with FA and CC. In composite, resistivity has increased with increasing proportion of CC from 10 to 40%. Beyond 40% i.e., 50% CC, there is a decline of resistivity. Thus, optimum amount of CC is 40% and has yielded best resistivity. Increasing resistivity is due to formation of higher amounts mullite in the sintered composite product. From the results, it concludes that FA60/CC40 composite is the best combination for preparing an insulator [18].

TGA runs are taken only for optimized composite, i.e., FA60:CC40 composite (Figure 4.6). For this, weight loss is estimated for different temperature, ranging between 30-800 °C (vide-Figure 4.6). There is a two stage sharp decrease in weight percentage below 360 °C. This is attributed to loss of water molecules from crystalline mullite phases [19, 20].

Frequency dependence dielectric constant of the composite (FA60:CC40) (k) is measured and is plotted (Figure 4.7). Composite material shows negative dielectric response both for low and high frequency ranges. High dielectric response ($k > 100$) at low frequencies suggests that there is a polarization of space charge [21]. As it is very desirable to obtain a high dielectric response at higher frequencies, it is also interesting to note that at much higher frequencies (>100 kHz) the dielectric response of composite is more negative. This behavior may originate from a suppressed space charge in the composite system. Dielectric constant of FA60:CC40 increases in presence K^+ and Na^+ cations and decreases when they are replaced by Ca^{2+}, Mg^{2+}, and Ba^{2+} cations. Dielectric property is determined

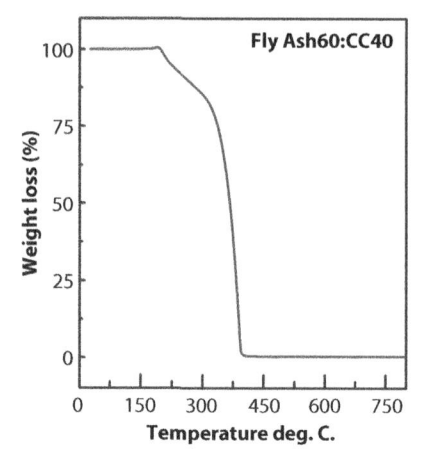

Figure 4.6 TGA plot of fly ash/kaolin composite (60:40) [15].

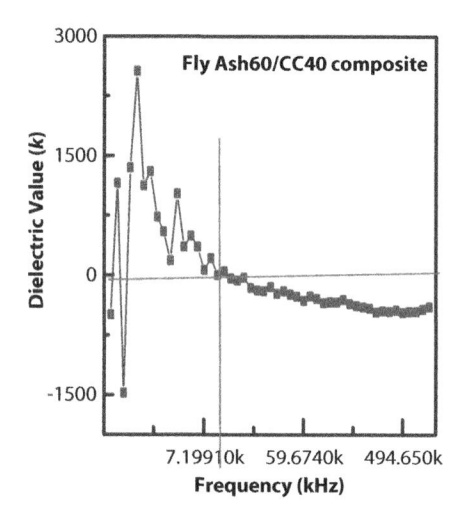

Figure 4.7 Dielectric plot of fly ash/kaolin composite (C) [15].

by the concentration and mobility of K^+ and or Na^+ ions in this phase. On the other hand, mullite phase has a vital role on dielectric properties. XRD analyses indicated that the composite shows higher percentage of crystalline phases, which can explain the property. Sample shows a particularly pronounced effect and may serve as a new avenue toward materials for electromagnetic cloaking and extremely low loss communications applications [21–24].

4.6 Conclusions

Electrical insulating materials are prepared with fly ash and kaolin.

Maximum resistance of 39.5 M.Ω is found for optimum combination at sintering temperature 1200 C for 1 h.

Microstructures of electrical insulating materials show compact mass consisting of amorphous and crystalline phases.

FESEM/EDS analyses show presence of elements such as Si, Al, O, etc.

XRD pattern show clearly formation of mullite and quartz phases in sintered products.

Numbers of peaks of sintered samples are due to formation of new bonds within the phases.

TGA curves have revealed for FA/CC composite, there is removal of water i.e., moisture from the material in the temperature range of 200-360 °C due to desorption of occluded water.

Within the frequency range i.e., 7.2 kHz to 500 kHz dielectric constant is observed to be negative, which is due to dielectric relaxation.

Acknowledgements

First author would like to thank Prof. Munesh Chandra Adhikary, PG Council Chairman, Fokir Mohan University, for his invaluable guidance, advice, and constant inspiration throughout the entire program. First author would also like to thank Mr. Mukteswar Mohapatra in Fokir Mohan University for his support. Authors convey their sincere thanks to GIET, University Gunupur, Rayagada, Odisha, India for providing Lab facilities to do the research work. Authors also like to thank the CRF, IIT Kharagpur for providing their testing facilities.

References

1. N. S. Pandian, C. Rajasekhar, and A. Sridharan, (1998), Studies of the Specific Gravity of Some Indian Coal Ashes, *Journal of Testing and Evaluation*, Vol. 26, p. 177.
2. N. S. Pandian, (2004), Fly Ash Characterization with Reference to Geotechnical Applications, *Journal of the Indian Institute of Science*, Vol. 84, pp. 189.
3. M. Ahmaruzzaman, (2010), A Review on the Utilization of Fly Ash, *Progress in Energy and Combustion Science*, Vol. 36, pp. 327.

4. L. Wang, G. Hu, F. Lyu, T. Yue, H. Tang, H. Han, Y. Yang, R. Liu, and W. Sun, (2019), Application of Red Mud in Waste water treatment, *Minerals*, Vol. 9, pp. 281.
5. S. Sushil and V. S. Batra, (2008), Catalytic Applications of Red Mud, an Aluminium Industry Waste: A Review, *Applied Catalysis B: Environmental*, Vol. 81, pp. 64.
6. C. L. Goldsmith, A. Malczcwski, J. J. Yao, S. Chen, J. Ehmk, and D. H. Hinzel, (1999), RFMEMS-Based Tunable Filters, *International Journal of RF and Microwave Computer-Aided Engineering*, Vol. 9, pp. 362.
7. G. Subramanyam, F. V. Keuls and F. A. Miranda, (1998), Novel K-band Tunable Microstrip v-band Pass Filter using Thin Film HTS/Ferroelectric/Dielectric Multilayer Configuration, *IEEE Microwave Guided Wave Letter*, Vol. 8, pp. 78.
8. A. Tombak, J. P. Maria, F. T. Ayguavives, Z. Jin, G. T. Stauf, A. I. Kingon, A. Mortazawi, (2003), Voltage-Controlled RF Filters Employing Thin Film Barium-Strontium Titanate Tunable Capacitors, *IEEE Trans. Microwave Theory and Technology*, Vol. 51, pp. 462.
9. https://www.syalons.com/2019/01/08/ceramic-materials-in-defence-applications/
10. R. Kulke, G. Mollenbeck, C. Gunner, P. Uhlig, K. H. Drue, S. Humbla, J. Muller, R. Stephan, D. Stopel, J.F. Trabert, G. Vogt, M. A. Hein, A. Molke, T. Baras, A. F. Jacob, D. Schwanke, J. Pohlner, A. Schwarz, and G. Reppe, (2009), Ceramic Microwave Circuits for Satellite Communication, *Journal of Microelectronics and Electronic Packaging*, Vol. 6, pp. 27-31.
11. L. Y. V. Chen, R. Forse, D. Chase, and R. A. York, (2004), Analog Tunable Matching Network using Integrated Thin-Film BST Capacitors, *IEEE Microwave Theory and Technology Society*, Vol. 1, pp. 261.
12. M. Pu, L. Liu, W. Xue, Y. Ding, H. Ou, K. Yvind, and J. M. Hvam, (2010), Widely tunable microwave phase shifter based on silicon-on-insulator dual-microring resonator, *Optics Express*, Vol. 18, pp. 6172-6182 (https://doi.org/10.1364/OE.18.006172).
13. J. Nath, D. Ghosh, J. P. Maria, A. I. Kingon, W. Fathelbab, D. Paul, F. Z. Michael, and B. Steer, (2005), An Electronically Tunable Microstrip Band Pass Filter using Thin-Film Barium-Strontium Titanate (BST). *IEEE Trans. Microwave Theory and Technology*, Vol. 53, pp. 2707.
14. F. H. Wee1, F. Malek, S. Sreekantan, A. U. Al-Amani, F. Ghani1, and K. Y. You, (2011), Investigation of the Characteristics of Barium Strontium Titanate (BST) Dielectric Resonator Ceramic Loaded on Array Antennas, *Progress in Electromagnetic Research*, Vol. 121, pp. 181.
15. M.K. Panigrahi, (2021), Investigation of Structural, Morphological, Resistivity of Novel Electrical Insulator: Industrial Wastes, *Bulletin of Scientific Research*, Vol. 3, pp. 51-58.
16. B. J. Saikia, G. K Rao, and P. Sarathy, (2010), Fourier Transform Infrared Spectroscopic Characterization of Kaolinite from Assam and Meghalaya,

North Eastern India, *International Journal of Modern Physics*, Vol. 1, pp. 206-210.
17. R. P. dos Santos, J. Martins, C. Gadelha, B. Cavada, A. V. Albertini, F. Arruda, M. Vasconcelos, E. Teixeira, F. Alves, J. L. Filho, V. Freire, (2014), Coal Fly Ash Ceramics: Preparation, Characterization, and Use in the Hydrolysis of Sucrose, *Scientific World Journal*, Vol. 2014, pp.1-8.
18. M. K. Panigrahi, P. Kumar, B. Barik, D. Behera, S. K. Mohapatra, H. Jha, (2016), Frequency Dependency of Developed Dielectric Material from Fly Ash: An Industrial Waste, *In: 20th National Conference on Nonferrous Minerals and Metals*, 8-9th July 2016; Eds. Rakesh Kumar, K.K.Sahu & Abhilash, 143.
19. H. Wang, C. Li, Z. Peng, and S. Zhang, (2011), Characterization and Z., *Journal of Thermal Analysis and Calorimetry*, Vol. 105, 157-160.
20. D. L. Y. Kong and J. G. Sanjayan, (2010), Effect of Elevated Temperatures on Geopolymer Paste, Mortar and Concrete, *Cement and Concrete Research*, Vol. 40, 334-339.
21. P. Barber, S. Balasubramanian, Y. Anguchamy, S. Gong, A. Wibowo, H. Gao, H. J. Ploehn, and H.-C. Zur Loye, (2009), Review Polymer Composite and Nanocomposite Dielectric Materials for Pulse Power Energy Storage, *Materials*, Vol. 2, pp. 1697-1733.
22. N. Singh Mehta, P. K. Sahu, P. Tripathi, R. P. Manas, and R. Majhi, (2018), Influence of Alumina and Silica Addition on the Physico-Mechanical and Dielectric Behavior of Ceramic Porcelain Insulator at High Sintering Temperature, *Boletín de la sociedad española de cerámica y vidrio*, Vol. 57, pp.151-159.
23. Y. Lv, M. Rafiq, C. Li, and B. Shan, (2017), Study of Dielectric Breakdown Performance of Transformer Oil Based Magnetic Nanofluids, *Energies*, Vol. 10, pp. 1-21.
24. K. Belhouchet, A. Bayadi, H. Belhouchet, M. Romero, (2019), Improvement of Mechanical and Dielectric Properties of Porcelain Insulators using Economic Raw Materials, *Boletín de la Sociedad Española de Cerámica y Vidrio*, Vol. 58, pp. 28-37.

5

High Resistance Pond Ash Geopolymer Ceramics

Muktikanta Panigrahi[1]*, Ratan Indu Ganguly[2] and Radha Raman Dash[3]

[1]*Department of Materials Science, Maharaja Sriram Chandra BhanjaDeo University, Balasore, Odisha, India*
[2]*Department of Metallurgical Engineering, National Institute of Technology, Raurkela, Odisha, India*
[3]*CSIR-National Metallurgical Laboratory (NML), Jamshedpur, Jharkhand, India*

Abstract

Development of cementious materials with high compressive strength is one of the present challenges for R & D Scientists/Engineers. Herein, we report a solid state approach (through chemical activation) for synthesizing geopolymer from pond ash as a prime raw material. Pond ash based Geopolymer has shown compressive strength of 19 MPa comparable to ordinary Porland cement pastes. Thus, these materials are equivalent to 20 M grade of cement and can minimize the use of Portland cement (PC). Prepared geopolymer has been characterized by XRD, HRTEM, FESEM, FTIR, BET (for surface area) and resistivity. Structures of pond ash and PA based geopolymer are examined using XRD, HRTEM, and FESEM. BET surface area (and pore volume) of pond ash and PA based geopolymer are determined by N_2 adsorption-desorption isotherm at 77 K. Values are found to be 0.781 m^2 g^{-1} (0.003 cm^3 g^{-1}) and 2.123 m^2 g^{-1} (0.007 cm^3 g^{-1}), respectively. Resistance of Geopolymer is determined by two probe method. Strength properties and structural integrity of developed geopolymer demonstrate that it can be a promising material for practical application in constructional sectors.

Keywords: Pond ash, geopolymer, FESEM, HRTEM, electrical resistivity, BET surface area and pore volume

*Corresponding author: muktikanta2@gmail.com

5.1 Introduction

Coal-based thermal power stations contribute to pollution significantly. It is claimed that 750 million tonnes of fly ash is generated annually by thermal power stations [1]. Different waste materials are generated due to coal combustion. Generated pollutants are fly ash, bottom ash, pond ash, slags and flue gas. Fly ash is extracted from flue gas by passing through electrostatic precipitator or cyclone separator. This also becomes a source for re-generation of valuable metals and can be used as a source material for production of ceramics, zeolites, adsorbents and geopolymers [2–5].

Geopolymers are called inorganic polymers, which are generated through alkali-activation of aluminosilicate. They have adequate strength at ambient temperatures. They are environmental-friendly and can be adopted as suitable structural materials [6, 7]. Bottom ash is collected from bottom of boilers. They differ from fly ash in terms of particle size range [8]. Geopolymers are prepared from fly ash and looking to the similarity of the composition, it is thought to replace fly ash with bottom ash for preparation of geopolymer [9–12]. Depending on the composition of the coal source, both fly ash and bottom ash may contain heavy metals. In their dry state, they can cause health hazards. Therefore, bottom ash is stored under water. The resulting pond ash may be bottom ash mixed with some amount of fly ash settled under the water in ponds. Pond ash gets mixed with soil under land. Since ponds are fed with highly pollutants and therefore, it becomes a threatening proposition for human life. It is, thus, desirable to find a method for utilization of pond ash other than its storage in ponds.

Bottom ash can be collected either in dry state or in the wet state. Even if it is collected in wet state from pond, it has significant proportion of fly ash after remaining in the wet state for long periods of time.

Properties of pond ash depend on composition of coal used, burning conditions, time spent in the slurry pond and particle separation during wet storage [13]. For this reason, a few studies are made on the application of pond ash, including production of geopolymers [14, 15]. Reactivity of pond ash depends on size fraction. Fine fraction shows pozzolanic activity and can be used in the production of cement and concrete [16]. Coarser fraction of pond ash shows weak pozzolanic activity. This has led to beneficiation using mineral processing techniques to separate the size fractions for possible use as a lightweight aggregate [17].

Lee *et al.* have used pond ash to produce geopolymer pastes having reasonable strength. They have collected pond ashes from South Korean pond. They have used pond ash for the above work after removal of carbon and drying it without any size fractionation [14, 15].

All the above work has not considered the change in the physical properties of pond ashes after storage under wet conditions for a long time. After prolonged storage under water, soluble components of ashes are dissolved, making the material more porous with decrease of density. Upon treatment of this material with alkali for production of geopolymers, pond ash absorbs large amount of liquid phase, increasing the liquid to solid ratio of the mixture. This weakens mechanical property of geopolymers.

BET surface areas of pond ash go up to 25 times higher and thereby pore volumes increases 10 times greater than those of fly ashes from the same power plant [18]. Hence, it is suggested that ashes submerged in ponds have undergone chemical and/or physical changes by leaching out of soluble phases and hydroxylation [18]. Such chemical and physical changes will also depend on how long ashes have been stored under water. Thus, physical and chemical changes of pond ash are expected to depend on the duration of storage under water.

Reaction of pond ash to alkali solution is influenced by particle size. Grinding will help further to reduce particle size. Finer particle size enables higher reactivity of reactants promoting formation of Geopolymer [19]. Mechanical activation of pond ash may pave an effective way of utilization of pond ash for ormation of Geopolymer. However, some pond ash has presence of significant amounts of heavy metals i.e., As, Pb and Cr. Hence, pond ash containing heavy metals are restricted for use as raw materials for production of geopolymer [20].

S.K.Saxena *et al.* [21] have developed innovative strategies to create green concrete with improved strength properties and durability. They have prepared geopolymer cement by activating Pond ash with alkaline solution (14 M NaOH and sodium silicate solutions). Natural Ennore sand is used to prepare Geopolymer mortar. Alccofine powder is added during Geopolymerization process which has helped to increase compressive strength of the motar. Silica fume is also used to prepare geopolymer mortar. Curing at different temperatures has been done either in an electric oven or in microwave oven. It is observed that microwave oven curing has enabled better compressive strength with shorter curing time. Powder X-ray diffraction, heat evolution profile, TG studies, compression strength, and SEM are used to characterize the geopolymer materials. Durability test in sulphuric acid is considered to be essential. Durability results have shown better stability with microwave cured material in comparison to other methods of curing.

Muhamed Khodr et al. [22] have made studies comprehensively on Geopolymer and Geopolymer mortar. They have used brown coal fly ash obtained from two separate locations of an Australian power plant. They have obtained good compressive strength of their prepared material.

Byoungkwan Kim et al. [23] have used pond ash to prepare geopolymers. They have either not adopted any purification process or a minimal purification process. Synthetic basalt is used as foaming agent in the preparation of geopolymer. They have observed the highest compressive strength (26 MPa) of geopolymers after 7-days of curing at ambient temperature. The compressive strength (80MPa) is thus, enhanced considerably by grinding and sieving of pond ash.

Sunil kumar Saxena et al. [24] have used pond ash containing higher amount of heavy metals to prepare geopolymer cement. Thus, they have utilized pond ash fruitfully to prepare geopolymer cement by treating it with alkaline solution followed by thermal curing. Experimental studies are done to assess the activating influences of alkali combination i.e., alkali hydroxide and silicate ($NaOH/Na_2SiO_3$, $NaOH/Li_2SiO_3$, KOH/Na_2SiO_3 and KOH/Li_2SiO_3) on the mechanical properties of mortars.

Sujeong Lee et al. [25] have characterized pores in geopolymer with the help of X-ray tomography, mercury Porosimetry, and gas adsorption method. They observed Irregular geometry of pores and approximate equivalent perimeter diameter ranging between 20 to 60 nm equivalents. Within the volume of 0.00748 μm³, Porosity is determined to be 7.15%. Use of electron tomography is considered to be an important method for measuring the porosity and pore connectivity in geopolymers. Knowledge of porosity helps to co-relate structure with properties and enable to predict the durability and properties.

Foregoing reviews indicate pond ash can be used as a raw material for preparation of geopolymer (GP). However, pond ash differs in compositions since thermal power plants uses coal obtained from different collieries. In addition, pond ash used in present investigation undergoes metamorphism due to preservation under pond water for a longer time. It is thought interesting to study characteristics of Geopolymer developed from pond ash dumped from NALCO in the pond. Present investigation has shown, geopolymer can be made successfully and therefore, present paper has described preparation, microstructure, resistivity, surface area (pore volume) i.e., formation of geopolymer using pond ash collected from NALCO, which is located in Odisha state, India.

5.2 Experimental Details

5.2.1 Materials and Chemicals

Pond ash is a waste product obtained from burning of coal in boilers (Figure 5.1). Pond ash is procured from NALCO plant, Damanjodi, Odisha, India. Chemical composition of pond ash is given in Table 5.1. Silica (SiO_2) and alumina (Al_2O_3) are present in major percentages, which are expected to take part in the polymerization process. Presence of alkali compounds such as CaO and MgO help in the process of polymerization.

Aqueous solution of pure sodium silicate (Na_2SiO_3) has been used to react with pond ash. Sodium silicate is however, activated by mixing with appropriate amounts of sodium hydroxide (NaOH) [12 mL SS + 3 mL SH] which will react with pond ash to form Geopolymer.

Sika is a water soluble plasticizer and is bought from Vizag Market, Visakhapatnam, Andra Pradesh, India.

Synthesis of GEOPOLYMER is carried-out by solid state route via chemical process. Geopolymer is prepared using different components such as pond ash, alkali activated solution and SiKA (Figure 5.1). The process is carried out in four consecutive steps *i.e.*, Grinding, Mixing, Ramming, and Curing in a hot air oven, which is indicated in Figure 5.1 [26].

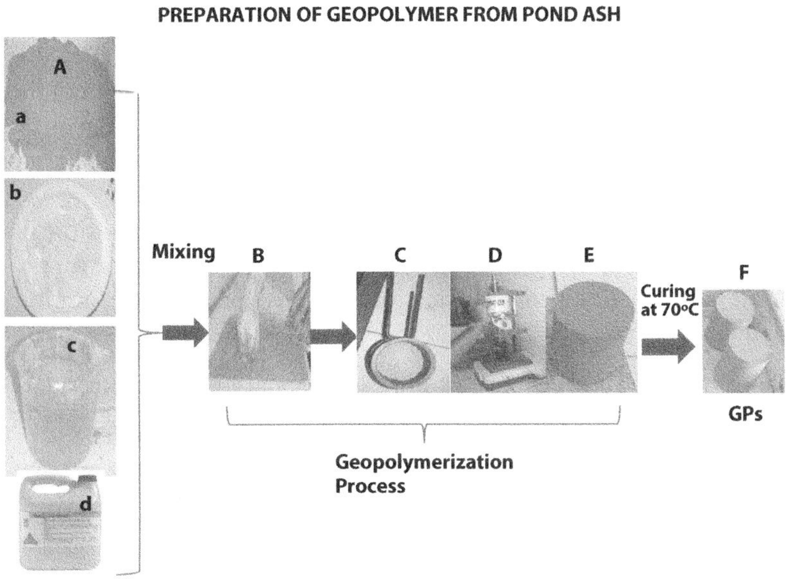

Figure 5.1 Schematic diagram of preparation process of GP from pond ash.

84 High Electrical Resistance Ceramics

Table 5.1 Chemical composition of pond ash (PA) in percentage (%) [26, 27].

Raw materials	SiO_2	Al_2O_3	CaO	MgO	Fe_2O_3	TiO_2	Cr_2O_3	MnO	P_2O_5	C	LOI
Pond Ash	62.8	28.3	0.7	0.58	3.85	1.84	0.04	0.03	0.32	1.15	0.5

Figure 5.1 describes schematic diagram of preparation of Geopolymer. Vertical column 'A' indicates raw materials (pond ash, NaOH, sodium silicate solution, and SiKA). Fine pond ash i.e., 240 meshes (Figure 5.1Aa) is used to prepare Geopolymer [26]. It is grounded by ball mill followed by sieving to 240 meshes. Sodium hydroxide (Figure 5.1Ab) is used to prepare 8 molar (M) alkali solutions. It helps to activate the polymerization process. Sodium silicate (Figure 5.1Ac) and water soluble plasticizers (Figure 5.1Ad) are used.

Alkali activated solution is prepared by taking optimum treatment combinations of sodium silicate (12 mL), sodium hydroxide (8 M, 3mL), SiKA (1-2 mL).

Horizontal diagram indicate different stages of geopolymer process. Figure 5.1B indicates mixing of raw materials (indicated vertical column Figure 5.1A) in appropriate proportion. A gel-like mass is formed.

Appropriate amount of gel-like masses are transferred to a REMI iron mould (having dimensions of 50 mm diameter and 70 mm height) and is shown in Figure 5.1C.

Figure 5.1D shows ramming (is a mechanical hammer) of mixed products. Samples are rammed continuously twenty (20) times to have dense compaction (Figure 5.1D).

Figure 5.1E shows cast green samples. It is obtained after demoulding.

Cast green sample (vide Figure 5.1F) is cured at 70 °C for 24 h in an oven.

The cured samples are preserved in plastic zipper bags until the samples are tested.

Three samples designated as S1, S2, S3 having same compositions are prepared (Table 5.2) in order to take average response of the mixtures.

Table 5.2 describes composition of mixture used for preparation of Geopolymer.

Table 5.2 PA based geopolymer mixtures (Three samples of same compositions).

Sample code	PA	Sod. silicate (SS)	Alkali (8 M, NaOH) (SH)	Water soluble plasticizer (Sika)
S1	85%	12%	3%	1-2ml
S2	85%	12%	3%	1-2ml
S3	85%	12%	3%	1-2ml

5.3 Test Methods

PA and PA-based GP materials are characterized by X-ray diffraction methods. For this, Phillips PW-1710 advance wide angle X-ray diffractometer is used. Powder samples are placed on sample holder, which is made from quartz. Samples are scanned from 10° to 70° diffraction angle (2θ) with scanning rate 2°/min.

FTIR spectra of PA and PA based GP are recorded on a Thermo Nicolt Nexus 870 spectrophotometer (400-4000cm^{-1}). Potassium Bromide (KBr) with small amounts of test powder samples are used to prepare pellet. Background spectrum is collected before running the samples.

Surface morphology and EDS of PA and PA based GP are obtained by field emission scanning electron microscope (Carl Zeiss Supra 40). Gold coating is done on sample via sputtering technique before placing into the instrument.

HRTEM instrument is used to investigate topology and microstructure of prepared GP and as-received PA. HRTEM instrument (JEM-2100 HRTEM, JEOL, and Japan) is operated at 200 kV acceleration voltages. Specimens (GP and raw PA) are prepared by microtone technique (LEICA Microsystem, GmBH, A-1170). Test samples are transferred to Cu TEM grids.

Effective Brunauer–Emmett–Teller (BET) surface area of processed pond ash and pond ash based Geopolymer samples are measured using Quanta chrome Chem BET analyzer. The experiment is done N_2 environment.

Compressive strength of cylindrical geopolymer samples (100 mm diameter and 200 mm depth) is measured. Testing procedure for compression strength is based on the AS 1012.9 (Standards Australia, 2014d). Compressive strength tests are conducted using the Controls MCC (Multifunctional Computerized Control Console) Machine (3000 kN capacity).

Electrical resistivity of pond ash based Geopolymer is measured by two probe method. Resistivity is determined by following equation [27];

$$\rho = \frac{R \times A}{l} \tag{5.1}$$

Here, A = area of electrode = $\frac{\pi}{4}D^2$ (m^2), R = resistance of prepared samples (MΩ), l = thickness of sample, ρ = estimated resistivity of the samples.

5.4 Results and Discussion

Optimum compositions and treatment combinations of variables are determined by statistical design of experiments which has been discussed elsewhere. Table 5.3 shows the average compressive strength of pond ash based Geopolymer with optimum treatment combinations. Maximum strength is found to be 19.0 MPa.

Figure 5.2 shows superimposed X-ray pattern of pond ash and treated pond ash. Main constituent phases are mullite and quartz [26]. It is interesting to observe that peak height of PA based geopolymer has increased with lesser base width. This is presumed to be due to crystallization of

Table 5.3 Compressive strength of PA based geopolymer (Treatment combinations-curing time (24 h), curing temperature (70 °C), Mesh 240) [26].

S. no.	Sample code	Pond ash	Sod. silicate (SS)	Alkali (8 M, NaOH) (SH)	Water soluble plasticizer (Sika)	Av. compressive strength (MPa)
1	`S1	85%	12%	3%	1-2ml	19

*Note: Results indicate average strength value obtained from 10 samples having similar compositions and same treatment combinations.

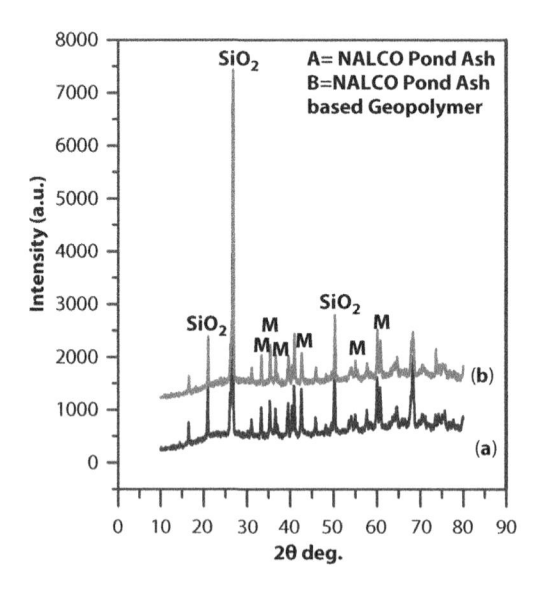

Figure 5.2 XRD patterns of: (a) As-received NALCO pond ash, (b) Pond ash based geopolymeric material [26].

glassy phases present in raw pond ash causing sharp increase in height and base width. As discussed earlier, Geopolymeric products are formed due to reaction between reactive glassy fraction of Pond Ash, sodium silicate, and NaOH with water soluble plasticizer. It is worthwhile to mention here that the reactive glasses participate in Geopolymerization.

Figure 5.3 show three HRTEM micrographs (Figure 5.3a, 5.3b, 5.3d) and diffraction pattern (Figure 5.3c) of raw material i.e., pond ash. Figure 5.3a and 5.3b are taken into two different magnifications as indicated [27]. Figure 5.3a (lower magnification) shows spherical and elliptical shaped particles spread in the background. Diameters vary 1 micron to 5 micron. At higher magnification (Figure 5.3b), some rod-shaped phases are found to be embedded in pond ash particles. The diameter is measured to be 0.5 nm to 100 nm. This signifies presence of multi-mineral phases in pond ash particles. SAD pattern of pondash (Figure 5.3c) indicates the presence of different (mineral) crystalline phases, which is corroborated in XRD pattern of pond ash (Figure 5.2).

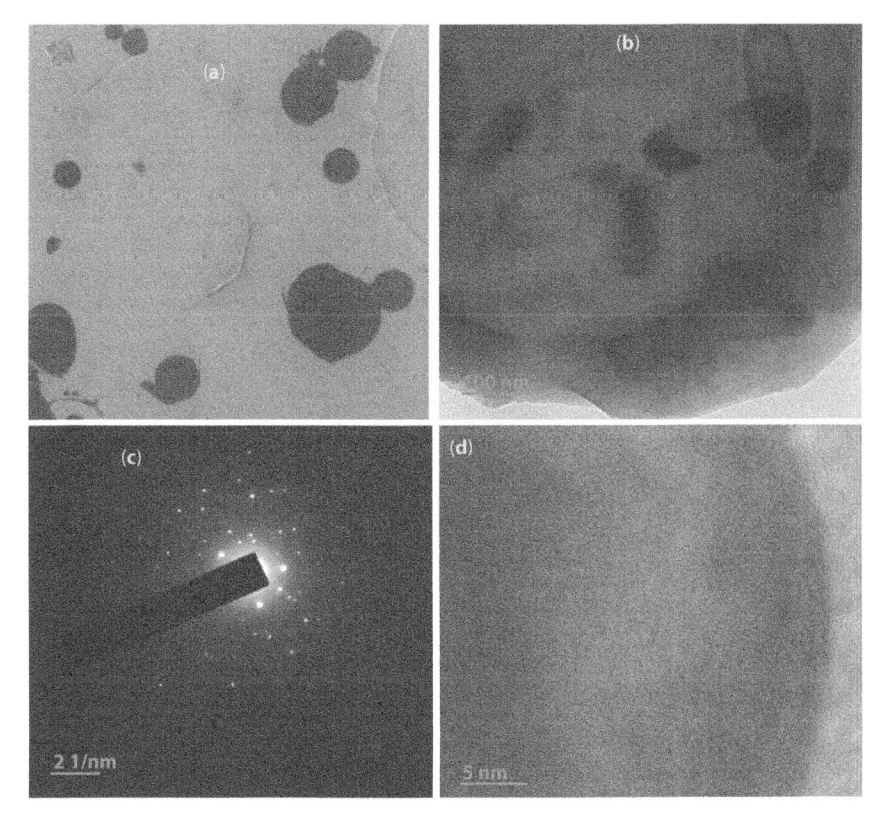

Figure 5.3 HRTEM image of PA (lower magnification, a), HRTEM image of PA (higher magnification, b), SAD pattern of PA (c), and Lattice pattern of PA (d).

88 HIGH ELECTRICAL RESISTANCE CERAMICS

Figure 5.4a and 5.4b indicate HRTEM micrograph of geopolymer prepared from Pond Ash. Figure 5.4a shows spherical particles fused together as a result of reaction occurring between raw materials i.e., pond ash and alkali materials. Figure 5.4b shows yet at higher magnification of pond ash based Geopolymer. Some porosity may be observed in the reacted products. Figure 5.4c depicts electron diffraction pattern taken from one of the spot which shows bright spots arranged in random manner. This indicates the crystallinity of product. If this pattern is compared Figure 5.3c, spots are fewer. Figure 5.4d indicates lattice pattern of pond ash based geopolymer. It depicts the presence of different crystalline lattice with orientation. Presence of crystalline phases is corroborated in XRD pattern of pond ash (Figure 5.2).

Figure 5.5 indicates FESEM micrographs of pond ash and pond ash based geopolymer. Figure 5.5a shows FESEM micrographs of Pond Ash.

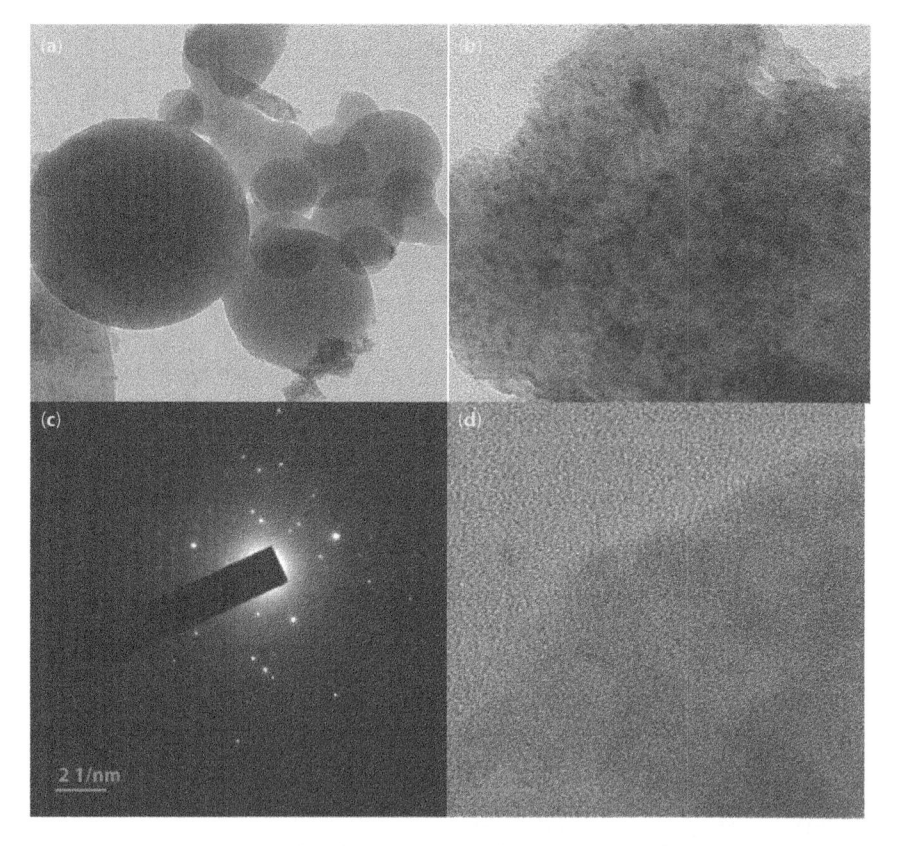

Figure 5.4 HRTEM image of PA based GP curried at 70 °C for 24h (lower magnification, a), HRTEM image of PA based GP curried at 70 °C for 24h (higher magnification, b), SAD pattern of PA based GP curried at 70 °C for 24h (c), and Lattice pattern of PA based GP curried at 70 °C for 24h (d).

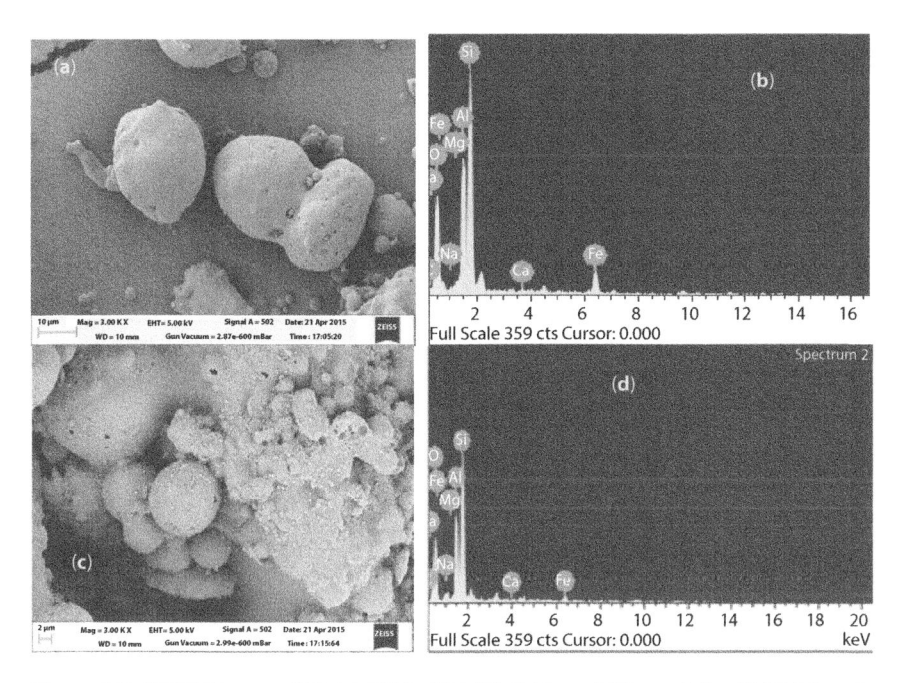

Figure 5.5 FESEM image of PA (a), EDS of PA (b), PA based GP curried at 70 °C (c), and EDS of PA based GP curried at 70 °C (d).

It shows irregular shaped porous particles. Corresponding EDS spectrum shows (Figure 5.5b) presence of elements such as Si, Al, Fe, Mg, Ca, Na, and O. Among the elements, Si, Al, and O are prominent.

Figure 5.5c depicts micrograph of Geopolymer formed due to reaction of pond ash with activated alkali materials. Figure 5.5c indicates compacted mass. Corresponding EDS graphs (Figure 5.5d) shows presence of elements such as Si, Al, Fe, Mg, Ca, Na, and O. Among the elements, Si, Al, and O are prominent.

Peak assignments with corroseponding wave number for FTIR spectra are shown in Table 5.4. Main feature of the FTIR spectra is the central band at around 1093 cm^{-1} and is attributed to the Si–O–Si or Al–O–Si asymmetric stretching mode. Main spectral band originally appeared for pond ash at 1078 cm^{-1} has shifted to lower frequency due Geopolymerization. Larger the shift, higher is the degree of penetration of Al from the glassy part of the pond ash to the $(SiO4)^{4-}$ net. This indicates that the Geopolymerization process is influenced by process parameters. Significant broad bands are observed at approximately 3450 and 1640 cm^{-1} due to O–H stretching mode and O-H bending mode.

Table 5.4 Wave number and peak assignment of as-received PA and PAGP [28, 29].

S. no.	Wave number		Peak assignments
	As-received PA	PA GP	
1	3450	3473	O-H stretching band
2	1625	1662	O-H bending band
3	---	2342	Carbonated band
4	---	1452	Asymmetric carbonate stretching
5	1093	1087	Si–O–Si or Al–O–Si asymmetric stretching
6	905	919	Mullite band
7	780	835	Quartz

Presence of quartz is shown by a characteristic band at around 835 cm^{-1}. Another spectral band at around 1452 cm^{-1} has appeared in the geopolymer sample, but is absent in pond ash. This band is characteristic of the asymmetric CO_3 stretching mode, which suggests presence of sodium carbonate as a result of the reaction between excess sodium and atmospheric carbon dioxide. Excess sodium content can form sodium carbonate by atmospheric carbonation and may disrupt the polymerization process. In the case of low-Ca alkali activated materials, CO_2 tends to form sodium carbonates and bicarbonates which is more soluble than the $CaCO_3$ formed in case of OPC. It will therefore act as an alkali sink and/or play a buffering role in the solution.

Specific surface area plays an important role on ceramic material's performance [30, 31]. Specific surface areas of as-received pond ash and PA based geopolymer are measured by N_2 adsorption-desorption isotherm at 77 K as shown in Figure 5.6. Both the samples show type IV isotherm as per Brunauer classification [30, 31]. BET surface area (and pore volume) of pond ash (Figure 5.6a) and PA based geopolymer (Figure 5.6b) are measured to be 0.781 m^2 g^{-1} (0.003 cm^3 g^{-1}) and 2.123 m^2 g^{-1} (0.007 cm^3 g^{-1}), respectively. Larger surface area of PA based geopolymer as compared to pond ash is attributed to the chemical changes occurring in the formation of geopolymer which has resulted in enhancement of active surface area [30, 31]. Therefore, in principle, it can offer more active sites for geopolymer reaction. At the same time, due to higher pore volume of GP (compared to pond ash), easy access of activated alkaline solution into pond ash surface can occur, giving rise to higher mechanical performance.

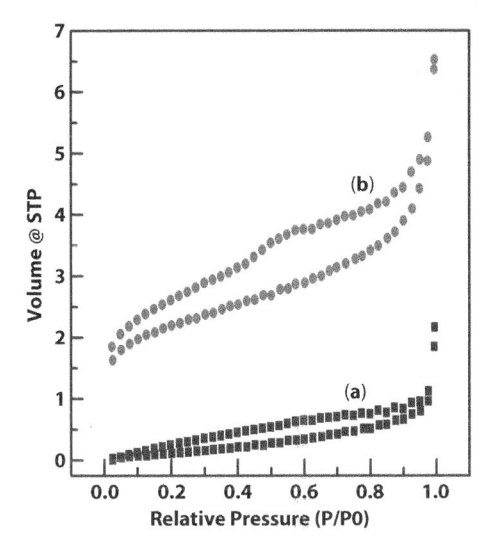

High Resistance Pond Ash Geopolymer Ceramics 91

Figure 5.6 Nitrogen adsorption/desorption isotherms of pond ash (a) and pond ah based geopolymer (b).

Table 5.5 indicates resistivity values of Pond Ash Geopolymers cured at 70 °C for different length of time. From Table 5.5, Pond Ash Geopolymers cured at 70 °C and 24 h shows the highest resistivity. Its resistivity value is measured to be 125.79705×10^6 Ω.cm. This happens due to progression of Geopolymerization process with increased curing time [32].

Table 5.5 Resistivity value of pond ash geopolymer materials [radius=1.5 cm and thickness=0.5 cm].

S. no.	Sample ID	Curing temperature (°C) and curing time	Resistance value (MΩ)	Resistivity value (Ω.cm)
1	Pond Ash	As received	---	---
2	Pond Ash Geopolymer	70 °C, 4 h	0.76	10.74222×10^6
3	Pond Ash Geopolymer	70 °C, 16 h	2.31	32.650695×10^6
4	Pond Ash Geopolymer	70 °C, 24 h	8.9	125.79705×10^6

5.5 Conclusions

Pond ash-based Geopolymer is successfully prepared by solid state route via chemical method using pond ash as a base component. Raw materials and GP are characterized with the help of FESEM/EDS, XRD, HRTEM, FTIR and BET surface area. The highest strength level achieved for Geopolymers is 19 MPa for a treatment combination of curing of 24 h and curing temperature of 70 °C. HRTEM analyses and FESEM micrographs reveal that there is a gradual transformation from irregular spherical shaped particles to compact porous mass. This is due to polymeric transformation for treatment combination of curing temperature 70 °C and curing time 24 h. Distinct changes are observed in FTIR spectra i.e, increase in peak height as well as appearance of many other peaks if compared with FTIR spectrum of virgin material. This strength is due to polymeric reactions and formation of chains with the monomeric structure. Strength level achieved by optimum use of variables is comparable to that of standard motar of grade (M15) as is used for constructional purpose. Maximum resistance is obtained for Geopolymer prepared with optimum treatment combination.

Acknowledgements

The authors convey their sincere thanks to the Ministry of Mines, Government of India for providing Grant to carry out the work [Grant number = F.No.:14/54/20214-Met.-IV and dated: 29.12.2014]. First author would like to thank Prof. Munesh Chandra Adhikary, PG Council Chairman, Fokir Mohan University, for his invaluable guidance, advice, and constant inspiration throughout the entire program. First author would also like to thank Mr. Mukteswar Mohapatra in Fokir Mohan University for his support. Authors convey their sincere thanks to GIET, University Gunupur, Rayagada, Odisha, India for providing Lab facilities to do the research work. Authors would also like to thank the CRF, IIT Kharagpur for providing their testing facilities.

References

1. A. Rastogi, and V. K. Paul, (2020), A Critical Review of the Potential for Fly Ash Utilization in Construction-Specific Applications in India, *Journal of Environmental Research, Engineering and Management*, Vol. 76, pp. 65-75.

2. G. A. Tochetto, L. Simão, D. de Oliveira, D. Hotza, and A. P. S. Immich, (2022), Porous Geopolymers as Dye Adsorbents: Review and Perspectives, *Journal of Cleaner Production*, Vol. 374, pp. 133982, https://doi.org/10.1016/j.jclepro.2022.133982.
3. S. Candamano, A. Policicchio, G. Conte R. Abarca C. Algieri S. Chakraborty, S. Curcio, V. Calabrò, F. Crea, and R. G. Agostino, (2022), Preparation of Foamed and Unfoamed Geopolymer/Nax Zeolite/Activated Carbon Composites for CO_2 Adsorption, *Journal of Cleaner Production*, Vol. 330, pp.129843, https://doi.org/10.1016/j.jclepro.2021.129843.
4. T. Samarina, E. Takaluoma, and O. Laatikainen, (2021), Geopolymers and Alkali-Activated Materials for Wastewater Treatment Applications and Valorization of Industrial Side Streams In book: *Advances in Geopolymers Synthesis and Characterization*.
5. P. Gupta, G. Nagpal, and N. Gupta, (2021), Fly Ash-based Geopolymers: An Emerging Sustainable Solution for Heavy Metal Remediation from Aqueous Medium, *Beni-Suef University Journal of Basic and Applied Sciences*, Vol. 10, pp. 1-29.
6. Nabila Shehata, O. A. Mohamed, E. T. Sayed, M. Ali Abdelkareem, and A. G. Olabi, (2022), Geopolymer concrete as green building materials: Recent applications, sustainable development and circular economy potentials, *Science of The Total Environment*, Vol. 836, p.155577, https://doi.org/10.1016/j.scitotenv.2022.155577.
7. M. Verma, N. Dev, I. Rahman, M. Nigam, M. Ahmed, and J. Mallick, (2022), Geopolymer Concrete: A Material for Sustainable Development in Indian Construction Industries, *Crystals*, Vol. 12, pp. 514-24, https://doi.org/10.3390/cryst12040514.
8. J. Yu, L. Sun, H. Tang, L. Jin, X. Song, S. Su, and J. Qiu, (2013), Physical and Chemical Characterization of Ashes from Municipal Solid Waste Incinerator in China, *Waste Management & Research*, Vol. 31, pp. 663-673.
9. J. Temuujin, A.Van Riessen, and K. J. D. MacKenzie, (2010), Preparation and Characterization of Fly Ash based Geopolymer Mortars, *Construction and Building Materials*, Vol. 24, pp. 1906-1910, https://doi.org/10.1016/j.conbuildmat.2010.04.012.
10. T. Bakharev, (2005), Geopolymeric Materials Prepared using Class F Fly Ash and Elevated Temperature Curing, *Cement and Concrete Research*, Vol. 35, pp. 1224-1232, https://doi.org/10.1016/j.cemconres.2004.06.031.
11. R. M. Kalombe, V. T. Ojumu, C. P. Eze, S. M. Nyale, J. Kevern, and L. F. Petrik, (2020), Fly Ash-Based Geopolymer Building Materials for Green and Sustainable Development, *Materials*, Vol. 13, pp. 5699-17.
12. N. B. Singh, (2018), Fly Ash-Based Geopolymer Binder: A Future Construction Material by Minerals, *Journal name*, Vol. 8, pp. 299, https://doi.org/10.3390/min8070299.

13. D. P. Mishra and S. K. Das, (2014), Comprehensive Characterization of Pond Ash and Pond Ash Slurries for Hydraulic Stowing in Underground Coal Mines, *Particulate Science And Technology*, Vol. 32, pp. 456-465.
14. M. Panigrahi, P. Sivakrishnan, P. K. Ran, R. R. Dash, R. I. Ganguly, (2016), Structural Transformation of Pond Ash Geopolymer: A Novel Construction Material in 20th National Conference on Non-Ferrous Minerals and Metals-2016.
15. A. Jose, M. R. Nivitha, J. Murali Krishnan, and R. G. Robinson, (2020), Characterization of cement stabilized pond ash using FTIR spectroscopy, *Construction and Building Materials*, Vol. 263, pp. 120136-13, https://doi.org/10.1016/j.conbuildmat.2020.120136.
16. V. Vv, S. Gowda, and R. V. Ranganath, (2020), An Investigation on Pozzolanicity of Mechanically Activated Pond Ash, *IOP Conference Series Materials Science and Engineering*, Vol. 936, 012004.
17. S.-J. Lee, H. Cho, and J. Kwon, (2012), Beneficiation of Coal Pond Ash by Physical Separation Techniques, *Journal of Environmental Management*, Vol. 104, pp. 77-84.
18. U. Bayarzul, D. S. Kim, S.-Ho. Lee, H. J. Lee, C. H. Ruescher, and K. J. D. MacKenzie, (2017), Properties of Geopolymer Binders Prepared from Milled Pond Ash, *Materiales de Construcción*, Vol. 67, e134.
19. B. Horvat and V. Ducman, (2020), Influence of Particle Size on Compressive Strength of Alkali Activated Refractory Materials, *Materials (Basel)*. Vol. 13, pp. 2227-16. PMCID: PMC7287946 (PMID: 32408672).
20. M. Romeekadevi, and K. Tamilmullai, (2015), Effective Utilization of Fly Ash and Pond Ash in High Strength Concrete, *International Journal of Engineering Research & Technology* (Special Issue – 2015), NCRTET-2015 Conference Proceedings, Vol. 3, pp. 1-7.
21. S. K. Saxena, M. K. Gautam, and N. B. Singh, (2018), Effect of Alccofine Powder on the Properties of Pond Fly Ash based Geopolymer Mortar under Different Conditions, *Environmental Technology & Innovation*, Vol. 9.
22. M. Khodr, David W. Law, C. Gunasekara, S. Setunge, and R. Brkljaca, (2019), Compressive Strength and Microstructure Evolution of Low Calcium Brown Coal Fly Ash-based Geopolymer, *Journal of Sustainable Cement-Based Materials*, Vol. 9, pp. 1-18.
23. K. Byoungkwan, L. Bokyeong, C. Chul-Min, and L. Sujeong, (2019), Compressive Strength Properties of Geopolymers from Pond Ash and Possibility of Utilization as Synthetic Basalt, *Journal of the Korean Ceramic Society*, Vol. 56, pp. 365-373, https://doi.org/10.4191/kcers.2019.56.4.03.
24. S. K. Saxena, M. K. Gautam, and N. B. Singh, (2017), Influence of Alkali Solutions on Properties of Pond Fly Ash-based Geopolymer Mortar Cured under Different Conditions, *Advances in Cement Research* 30(1):1-7.
25. S. Lee, H. TaeJou, A. Riessen, William D. A. Rickard, C.-M. Chon and N.-H. Kang, (2014), Three-Dimensional Quantification of Pore Structure in Coal Ash-based Geopolymer using Conventional Electron Tomography,

Construction and Building Materials, Vol. 52, pp 221-226, https://doi.org/10.1016/j.conbuildmat.2013.10.072.
26. M. K. Panigrahi, R. R. Dash, and R. I. Ganguly, (2018), Development of Novel Constructional Material From Industrial Solid Waste as Geopolymer for Future Engineers in *IOP Conference Series: Materials Science and Engineering,* Vol. 410, pp.1-12.
27. M.K. Panigrahi, (2021), Investigation of Structural, Morphological, Resistivity of Novel Electrical Insulator: Industrial Wastes, *Bulletin of Scientific Research,* Vol. 3, pp. 51-58.
28. P. Innocenzi, P. Falcaro, D. Grosso, and F. Babonneau, (2003) Order to Disorder Transitions and Evolution of Silica Structure in Self-Assembled Mesostructured Silica Films Studied through FTIR Spectroscopy, *Journal of Physical Chemistry B,* Vol. 107, pp. 4711-4717.
29. A, Palomo, M. W. Grutzeck, and M. T. Blanco, (1999), Alkali-Activated Fly Ashes A Cement for the Future, *Cement and Concrete Research,* Vol. 29, pp. 1323-1329.
30. G. Williams, B. Seger, and P. V. Kamat, (2008), TiO_2-Graphene Nanocomposites. UV-Assisted Photocatalytic Reduction of Graphene Oxide, *ACS Nano,* Vol. 2, pp. 1487–1491.
31. L. Hao, H. Song, L. Zhang, X. Wan, Y. Tang, and Y. Lv, (2012), SiO_2/Graphene Composite for Highly Selective Adsorption of Pb(II) Ion, *Journal of Colloid Interface Science,* Vol. 369, pp. 381-387.
32. J. Cai, J. Pan, X. Li, J. Tan, and J. Li, (2020), Electrical Resistivity of Fly Ash and Metakaolin-based Geopolymers, *Construction and Building Materials,* Vol. 234, pp. 117868-9, https://doi.org/10.1016/j.conbuildmat.2019.117868.

6

High Resistance Sintered Pond Ash Ceramics

Muktikanta Panigrahi[1]*, Ratan Indu Ganguly[2] and Radha Raman Dash[3]

[1]*Department of Materials Science, Maharaja Sriram Chandra Bhanja Deo University, Balasore, Odisha, India*
[2]*Department of Metallurgical Engineering, National Institute of Technology, Raurkela, Odisha, India*
[3]*CSIR-National Metallurgical Laboratory (NML), Jamshedpur, Jharkhand, India*

Abstract

Coal ash (i.e., fly ash/pond ash/bottom ash) is a waste material, obtained from thermal power plants. It causes environmental pollutions. It also creates disposal problems. Therefore, it is thought that sintering of this material will enable to develop value added products.

In the present work, sintering of pond ash (PA) is carried out at different temperatures (i.e., 900 °C, 1100 °C, and 1300 °C) for 1 h. Sintering of pond ash is carried in the form of pellets. Pellets are prepared at room temperature under compaction pressure of 10 MPa. Phases present in sintered materials are identified by X-ray diffraction (XRD) technique. Examination reveals presence of phases (i.e., quartz, mullite, etc) by XRD analyses. Microstructures of sintered products are examined by scanning electron microscope (SEM). Presence of mullite phase in sintered products has improved insulating properties. EDS analyses have detected presence of elements in sintered materials. Occurrence of different chemical groups of raw pond ash and sintered pond ash are made by FTIR analyses. Electrical resistivity of sintered materials is calculated by two-probe technique. The highest resistivity is found to be 1054.4337×10^6 Ω.cm.

Keywords: Industrial wastes, pond ash, XRD, phases, mullite, insulator, morphology, resistivity

---

*Corresponding author: muktikanta2@gmail.com

Muktikanta Panigrahi, Ratan Indu Ganguly and Radha Raman Dash. High Electrical Resistance Ceramics: Thermal Power Plants Waste Resources, (97–114) © 2023 Scrivener Publishing LLC

6.1 Introduction

In thermal power-plants, Coal ash is produced during combustion of coal [1–4]. Chemical and physical properties of coal ash depend on chemical composition and burning process [5]. Elements present in major quantities of coal ashes are silicon, calcium, aluminum, iron, magnesium, sulfur, carbon, etc. However, elements such as Pb, As, Sb, etc are present in traces. [6, 7].

Coal ash is classified in three groups i.e., fly ash, bottom ash, and pond ash [8]. Fly ash is considered to be by-products. Fly ash is collected from electrostatic precipitators, mechanical filters, or is simply left to atmosphere [9]. Fly ash is composed of hollow spherical particles and is mainly found in cenospheres along with smoke. Particles are low density material and are composed of glassy aluminum-silicate matrix in which quartz, mullite, calcite, magnetite are embedded [10].

In thermal power plants, generally, fly ashes are produced around 70% due to combustion of coal [11]. Occurrence of this residue in the environment is highly detrimental to our society as it pollutes environments (soil, water, and air) [12]. Based on their composition, fly ash has been potentially used as additive in many applications (*i.e.,* manufacture of cement [13] and concrete [13], conventional ceramics [14], and vitroceramics [15], production of tiles [16].

Pond ash is a waste product resulted from burning of coal in boilers. It is mainly obtained from wet disposal of fly ash, which when gets mixed with bottom ash is disposed in large pond or dykes as slurry. Pond ash is being generated in an alarming rate. Generation of pond ash is posing a threat to our environment and therefore, management is now experiencing problem to dispose of it. Therefore, it has become the thrust area for engineering research [14].

Usually, industrial wastes contain different types of elements (*i.e.,* Cu, Pb, Cd, Ag, Mo, Fe, Ti, Na, Mn, S, P, Zn, and Cl) in different amount. Some of these elements increase insulating and some of them enhance semiconducting property [15, 16]. Hence, one way is to explore possibility for using waste in making electrical materials (insulators and semiconducting) by suitable treatment of pond ash.

Dielectric properties and DC resistivity of industrial waste based materials have been studied at higher temperature (300-500 °C) [17].

Microstructure of ash mainly comprises of crystalline and glassy phases [18]. Work is going on to improve dielectric and electrical insulation properties from these industrial waste [19–22].

DC resistivity, dielectric constant, and dielectric loss are main considerations for fabricating insulator [23, 24].

Ceramics are used for different purposes *viz.*, refractories [25, 26], cutting tools [27], thermal insulations [28], etc. Ceramic components are prepared at high pressure and high temperature. Performances of the ceramics are assessed by presence of crystalline phase(s). Sintering process parameters i.e. temperature and time play an important role to achieve desired phases. Formation of mullite phase in the system is a key issue Mullite phase can be formed at firing temperature between 1150 to 1200 °C [29].

In early 1850, ceramic materials were used as electrical insulator for construction of electrical air lines [30]. Ceramics have number of distinctive properties (*i.e.,* mechanical strength, high-power dielectric strength, and corrosion resistance) [31, 32]. In current scenario, low cost insulator is needed for electrical engineering sectors. Current demand motivates researchers to develop insulators for meeting present requirements [33]. Insulators have great importance in our day-to-day life. It makes our life easy, harmless and shock free [34–36].

In the present work, pond ash-based materials are prepared through sintering process via solid state route. Materials are prepared at different sintering temperatures (i.e., 900 °C, 1100 °C, and 1300 °C). Structures of sintered materials are analyzed by SEM with EDS. Phases presents in sintered pond ash materials have been investigated by X-ray diffraction technique. FTIR is done to identify presence of different chemical groups in pond ash based sintered samples. Electrical resistivity value is also calculated.

6.2 Experimental Details

6.2.1 Materials and Chemicals

Pond ash is obtained on demand from National Aluminum Company, Damanjodi, Odisha, India. Composition of received pond ash is shown in Table 6.1. Dextrin $(C_6H_{10}O_5)_{n.}xH_2O$) is procured from Loba chemicals. This is used to bind pond ash particles. During preparation, 6% water is added to the above mixture (i.e. pond ash and dextrin). Optical images of pond ash and dextrin are shown in Figures 6.1a and 6.1b.

6.2.2 Materials Preparation

Sintering of Pond Ash is achieved through solid state route. Following steps are adopted to obtain PA based sintered materials [38].

Table 6.1 Presence of constituents with percentage (%) of NALCO pond ash [37].

	Used industrial waste
	Pond ash
Constituents with percentages	Fe_2O_3 = 3.85 MgO = 0.58 Al_2O_3 = 28.3 SiO_2 = 62.8 P_2O_5 = 0.32 Cr_2O_3 = 0.04 ZnO = 0.027 MgO = 0.049 C = 1.15 CaO = 0.7 TiO_2 = 1.84 MnO = 0.03 LOI = 0.5
Sources	NALCO, Damanjodi, Odisha, India
References	[37]

Figure 6.1 Optical image of pond ash (a) and dextrin (b).

Step-1 Preparation of raw materials (i.e., Pond ash)
As received pond ash is grinded in a ball mill for 5 h. It is sieved through 240 meshes. Sieved pond ash is dried in an oven at 120 °C (2 h). Purpose of drying is to remove adhered moisture from sieved pond ash.

Step-2 Preparation of green pellets
10 g of processed pond ash is kept in a mortar and is subsequently, mixed with small amount of dextrin (i.e., 0.05 %). Mixed product is grinded for

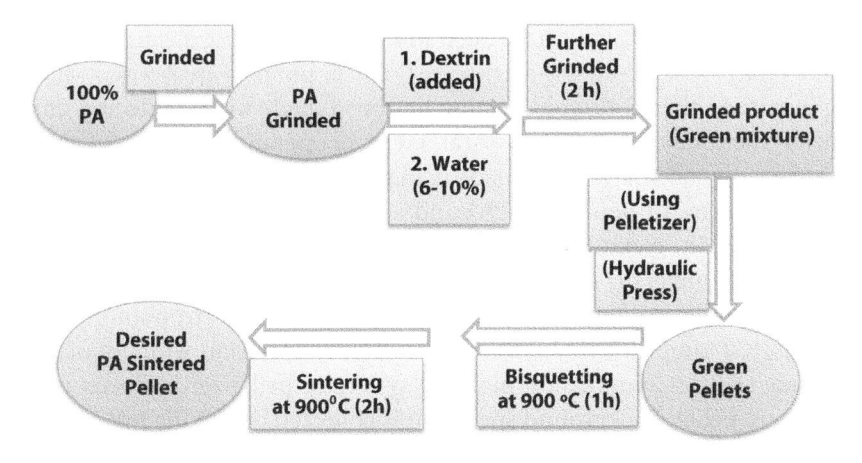

Figure 6.2 (Scheme 6.1) Flow chart for preparation of sintered pond ash based materials.

1h in the mortar and pastel. Small amount of water has been added to the mixture in order to get it in the form of paste. Samples are ready for preparing pellets.

Pellets are prepared in a pelletizer under pressure applied through universal testing machine (UTM). Compaction pressure of 10 MPa is maintained for five minutes. Then, pellets are removed from moulding system and are collected. These are green pellets. Green pellet is now ready for sintering process.

Step-3 Sintering of pond ash based pellet materials
Green pellet samples are kept in a muffle furnace. It is bisquetted at 900 °C for 1 h to remove intact water and binder. Bisquetted pellet(s) are further sintered at 900 °C/1100 °C/1300 °C for 2 h. Dimensions (i.e, thickness and diameter) of sintered pellet are found to be 0.5 cm and 3 cm, respectively. Description of above process has been shown schematically in Figure 6.2. Sintered pellet(s) is ready for different characterizations process.

6.2.3 Test Methods

X-ray diffraction (XRD) technique has been used to detect phases present in pond ash as well as in sintered pond ash. For XRD analyses, Phillips PW-1710 advance wide angle X-ray diffractometer are used. Pellet samples (un-sintered and sintered) are placed on a sample holder, separately. Samples are scanned through 20-80° diffraction angles with scanning rate of 2°/min.

Morphological analyses have been made with scanning electron microscope (Carl Zeiss Supra 40 instrument). Since, samples are non-conducting in nature; therefore, gold coating has been made with sputtering unit. During the test, an instrument parameter *i.e.,* operating voltage is kept at 30 kV. Elemental analyses are made using EDS attached to SEM.

FTIR spectra of pond ash (PA) and sintered PA are carried out using Thermo Nicolt Nexus 870 spectrophotometer (range 400-4000cm^{-1}). Settings parameters (*i.e.*, 50 scan at 4 cm^{-1} resolution, Absorbance measurement mode) are kept constant and are fixed in the spectrophotometers. For this, test samples are prepared by mixing raw materials with potassium bromide (KBr). Compression molding system in pelletizer has been used to prepare pellet. Background spectrum is collected before the test.

In electrical resistivity measurement, sintered pellet samples are used. Resistivity is estimated using the following relation [39].

$$\rho = \frac{R \times A}{l} \quad (6.1)$$

Where, "A" is the area of cross-section $= \frac{\pi}{4} D^2 (cm^2)$, "R" is the resistance of prepared samples in MΩ, "l" is thickness of the sample, "ρ" is the resistivity of the samples in MΩ.cm.

6.2.4 Results and Discussion

SEM image of unsintered pond ash is shown in Figure 6.3a. Figures 6.3b, 6.3c, and 6.3d show SEM image of sintered pond ash at three different temperatures (900 °C, 1100 °C, 1300 °C). Distinction can be made between unsintered and sintered materials. In unsintered pond ash (Figure 6.3a), there are spherical and elliptical particles scattered in the background. Particles have different dimensions varying between 1 to 8 micrometer.

In pond ash sintered at 900 °C for 1h, coagulation of particles is found to have occurred (Figure 6.3b). At higher sintering temperatures (Figure 6.3c and Figure 6.3d), there is incipient fusion and agglomerations of particles, resulting less voids in the matrix. Because of reduction of voids, phases appear to be glassy in nature.

Elemental distribution profile of pond ash and sintered pond ash are made with EDS attached to SEM. Figure 6.4a, 6.4b, 6.4c, and 6.4d depict elements present in both pond ash and sintered pond ash. In all the samples, elements (i.e., O, Mg, Ca, Al, Si, and Fe) are commonly present. Table 6.2 shows

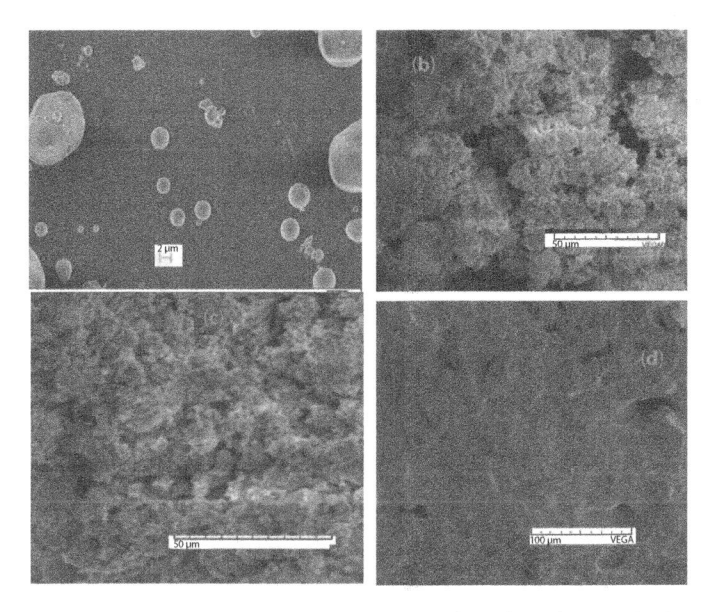

Figure 6.3 Morphological investigation of as-processed pond ash (a), pond ash sintered at 900 °C (b), pond ash sintered at 1100 °C (c), pond ash sintered at 1300 °C (d).

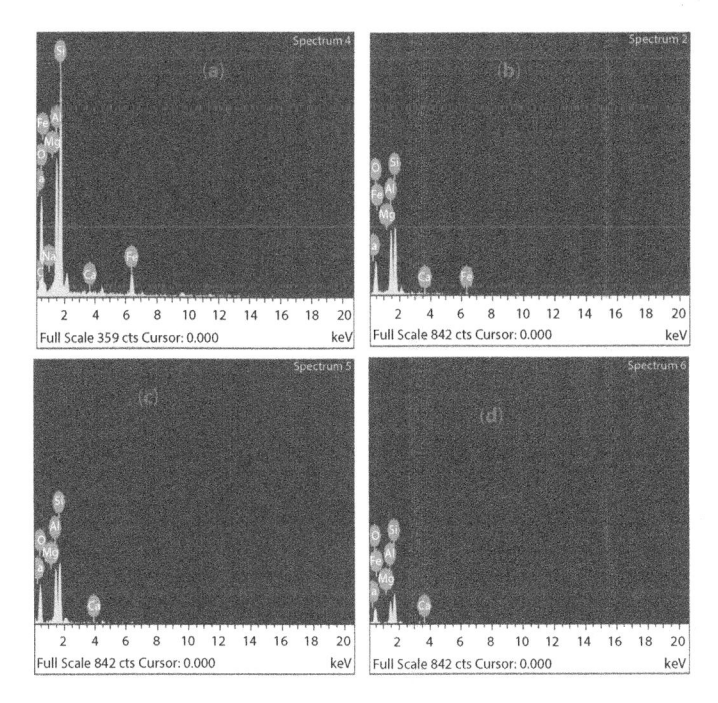

Figure 6.4 EDS spectrum of as-processed pond ash (a), sintered PA at 900 °C (b), sintered PA at 1100 °C (c), and sintered PA at 1300 °C (d).

Table 6.2 Elements present (with percentage) in EDS images (Figure 6.4).

S. no.	Elements (weight in %)	Sample ID			
		S1	S2	S3	S4
1	C	0	0	0	0
2	O	39.75	47.02	50.84	51.71
3	Na	0	0	0	0
4	Mg	0	0	0	0
5	Al	15.15	15.94	19.21	16.10
6	Si	33.32	34.13	29.96	32.18
7	Ca	0	0	0	0
8	Fe	11.78	2.89	0	0

Note: S1-as processed pond ash, S2-Pond ash sintered at 900 °C, S3-Pond ash sintered at 1100 °C, S4-Pond ash sintered at 1300 °C.

quantitative analyses i.e., relative amount of various elements present in pond ash and sintered pond ash (at 900 °C, 1100 °C, and 1300 °C).

X-ray diffraction patterns of pond ash and sintered pond ash are shown in Figure 6.5 (a, b, c, and d). In all cases, some differences can be observed. Principal phases observed are different. Table 6.3 has shown identified principal phases occurring for pond ash sintered at different temperatures. Occurrence of prominent phases of mullite and silicon dioxide are reported in the Table 6.3, which have occurred at higher sintering temperature (i.e., at 1300 °C). At lower sintering temperatures (i.e., 900 °C/1100 °C), phases identified are quartz, which is hexagonal in nature. It may be mentioned that there are many other lines present in the pattern, which may be due to the presence of other phases. Information reported in Table 6.3 is obtained from ICSD analyses. In all diffraction patterns, there are several lines, which may be due to presence of other phases i.e., silica, alumina, etc. The XRD pattern also reveals terbium oxide ($TbO2$) phase for pond ash sintered at 1100 °C.

FTIR absorption bands of as-received NALCO pond ash have been discussed elsewhere. Presence of different absorption stretching frequency of pond ash powder corresponds to its components. Stretching frequency at 600 cm^{-1} matches to Si-O-Al band. Stretching frequencies at 1098 and

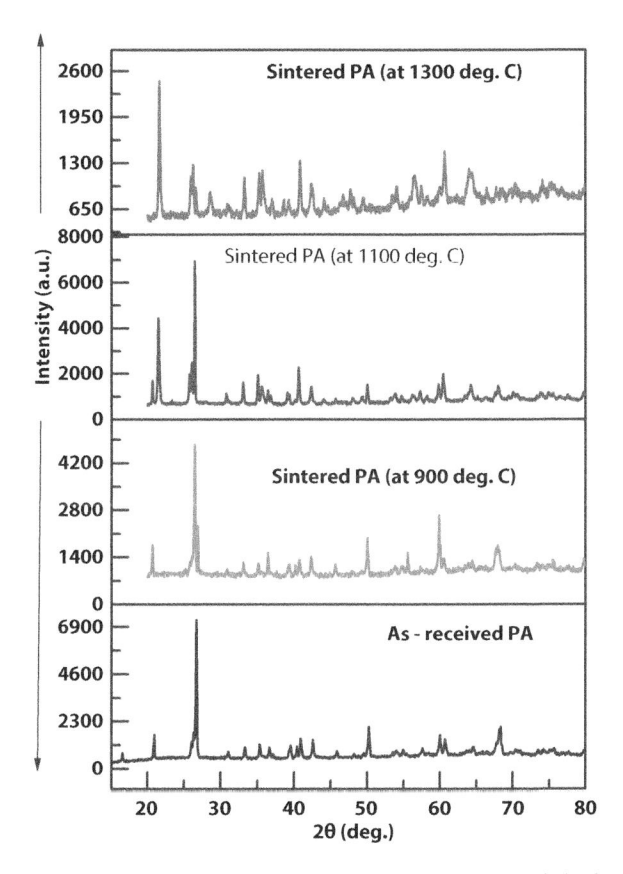

Figure 6.5 XRD pattern of unsintered pond ash and sintered pond ash (indicated in the figure).

1608 cm^{-1} may be accredited to Si-O-Si asymmetric band and H-O-H bending band, respectively [40–44].

FTIR spectrum of 1300 °C sintered pond ash is shown in Figure 6.6. Figure 6.6 indicates presence of different absorption bands occurring at wave numbers i.e., 3442, 2928, 1812, 1604, 977, and 878 cm^{-1}. Bands are in agreement with stretching vibrations of O-H bonds (3442 cm^{-1} wave number) and H-O-H bending vibrations (1604 cm^{-1} wave number) of interlayer adsorbed H$_2$O molecule. The hydroxyl-stretching band of water plays an important role and peak shift of FTIR spectra is significant. Absorption band ensue at 977 cm^{-1} wave number is attributed to Si-O band and signifies occurrence of silicate groups. Presences of Al^{3+}O^{2-} absorption bands are also indicated near 878 cm^{-1} wave number [40–44].

Table 6.3 Structural information of pond ash based samples.

Crystal ID	Mineral name	Crystal system	Space group	Space group number	Cell dimension	Crystal angle	Calculated density (g/cm^3)
Processed PA	Quartz	Hexagonal	P3221	154	a=b≠c	α=β=90°, γ=120°	2.65
PA 900 °C	Quartz (low)	Hexagonal	P3121	152	a=b≠c	α=β=90°, γ=120°	2.66
PA 1100 °C	Quartz	Hexagonal	P3221	154	a=b≠c	α=β=90°, γ=120°	15.90
PA 1300 °C	Mullite	Orthorhombic	Pbam	55	a≠b≠c	α=β=γ=90°	3.17
PA 1300 °C	Silicon dioxide	Cubic	Fd-3m	227	a=b=c	α=β=γ=90°	2.21

Note: PA- Pond ash.

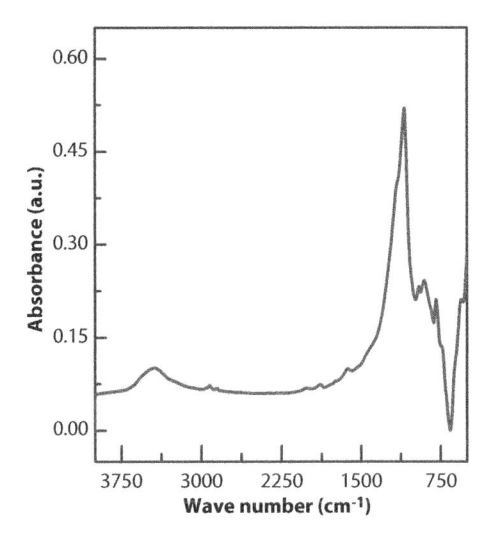

Figure 6.6 FTIR spectra of pond (sintered at 1300 °C).

Resistivity values of pond ash sintered at 900 °C, 1100 °C, and 1300 °C are presented in Table 6.4. From the estimated resistivity data, the highest resistivity is found to be 1054.4337 × 10^6 Ω.cm (for pond ash sintered at 1300 °C) and the lowest resistivity is found to be 13.427775 × 10^6 Ω.cm (for raw pond ash). Enhancement resistivity of sintered pond ash is due to formation of mullite phases in sintering of pond ash [45]. This is well corroborated with XRD pattern obtained for sintering pond ash if compared with as-received pond ash.

Table 6.4 Resistivity value of pond ash based sintered samples [radius=1.5 cm and thickness=0.5 cm].

S. no.	Sample ID	Sintering temperature (°C)	Resistance value (MΩ)	Resistivity value (Ω.cm)
1	S1 [22]	0	0.95	13.427775 × 10^6
2	S2	1300	74.6	1054.4337 × 10^6
3	S3	1100	49.1	694.00395 × 10^6
4	S4	900	23.7	334.98765 × 10^6

S1-as processed pond ash, S2-pond ash sintered at 1300 °C, S3-pond ash sintered at 1100 °C, S4-pond ash sintered at 900 °C.

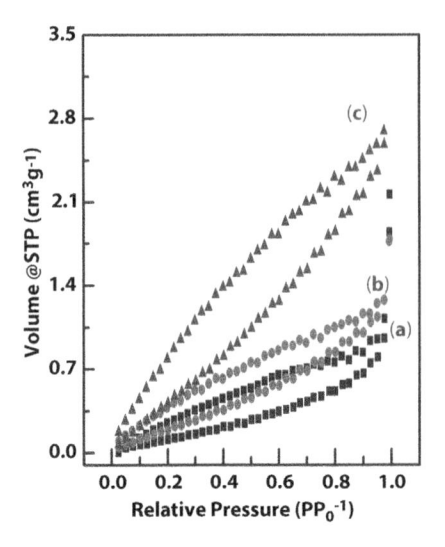

Figure 6.7 Nitrogen adsorption/desorption isotherms of pond ash (a) 1100 °C sintered PA (b), 1200 °C sintered PA (c).

Specific surface area plays an important role on performance of ceramic materials [30, 31].45 Figure 6.7 show variation of BET surface with relative pressure for as-received pond ash and sintered PA. It is measured with the help of N_2 adsorption-desorption isotherm at 77 K. Samples show type IV isotherm as per Brunauer classification [30, 31]. In all cases, BET surface areas have increased with increasing relative pressure. However, surface area increases rapidly with increasing relative pressure of 0.9. Measured specific surface area (and pore volume) of pond ash (Figure 6.7a), sintered PA at 1100 °C (Figure 6.7b), and sintered PA at 1300 °C (Figure 6.7c) are measured to be 0.781 $m^2\ g^{-1}$ (0.003 $cm^3\ g^{-1}$), 2.267 $m^2\ g^{-1}$ (0.003 $cm^3\ g^{-1}$), and 2.559 $m^2\ g^{-1}$ (0.007 $cm^3\ g^{-1}$), respectively. This is attributed sintering of materials [30, 31]. Therefore, in principle, it can offer more active sites for the pond ash reaction. Hence, more surface reaction can occur, giving rise to higher mechanical performance.

6.3 Conclusions

Pond ash is sintered at 900 °C, 1100 °C, and 1300 °C. Structure and morphology of the materials are studied using XRD and FESEM characterization techniques, respectively.

Characterization of materials indicates formation of mullite due to sintered at higher temperature. By comparing XRD pattern it is concluded that mullite formation is enhanced with increasing sintering temperature.

Different phases i.e., mullite and quartz phases are clearly observed in XRD pattern. Unit cell structures of mullite and quartz is orthorhombic and Tetragonal, respectively for pond ash sintered at 1300 °C. Estimated densities of orthorhombic and cubic cells are 3.17 g/cm^3 and 2.21 g/cm^3, respectively.

Microstructures reveal formation of coagulated mass at higher temperature of sintering due to incipient fusion of raw materials. Together with microstructural analyses, EDS analyses show presence of elements in sintered materials.

FTIR analysis reveals stretching bands at 3442, 1604, 977, and 878 cm^{-1} correspond to O-H bonds, H-O-H bending vibrations, Si-O band, and $Al^{3+}O^{2-}$ band.

BET surface areas of as-received pond ash and sintered materials are evaluated. It is observed that there is 3 to 3.5 fold enhancement of active surface area of pond ash sintered at 1300 °C.

There is marked increase in electrical resistivity of sintered materials, which is attributed to the formation of mullite. The highest electrical resistivity sintered pond Ash (1300 °C) is found to be 1054.4337 × 10^6 Ω.cm. It is therefore very much feasible to develop an insulator from waste i.e., pond ash.

Acknowledgements

First author would like to thank Prof. Munesh Chandra Adhikary, PG Council Chairman, Fokir Mohan University, for his invaluable guidance, advice, and constant inspiration throughout the entire program. First author would also like to thank Mr. Mukteswar Mohapatra in Fokir Mohan University for his support. Authors convey their sincere thanks to GIET, University Gunupur, Rayagada, Odisha, India for providing Lab facilities to do the research work. Authors also thank the CRF, IIT Kharagpur for providing their testing facilities.

References

1. S. Babel and T. A. Kurniawan, (2003), Low Cost Adsorbents for Heavy Metals Uptake from Contaminated Water: A Review, *Journal of Hazardous Materials*, Vol. 97, pp. 219-243.

2. P. S. Nayak and B. K. Singh, (2007), Instrumental Characterization of Clay by XRF, XRD and FTIR, *Bulletin of Materials Science*, Vol. 30, pp. 235-240.
3. L. Benco, D. Tunega, J. Hafner, and H. Lischka, (2001), Ab initio Density Functional Theory Applied to the Structure and Proton Dynamics of Clays, *Chemistry Physics Letter*, Vol. 333, pp. 479-484.
4. J. C.Hower, C. L. Senior, E. M. Suuberg, R. H.Hurt, J. L.Wilcox, and E. S. Olson, (2010), "Mercury Capture by Native Fly Ash Carbons in Coal-Fired Power Plants," *Progress in Energy and Combustion Science*, Vol. 36, pp. 510-529.
5. S. Singh, L. C. Ram, R. E. Masto, and S.K.Verma, (2011), Acomparative Evaluation of Minerals and Trace Elements in the Ashes from Lignite, Coal Refuse, and Biomass Fired Power Plants, *International Journal of Coal Geology*, Vol. 87, pp. 112-120.
6. H. B. Vuthaluru and D. French, (2008) Ash chemistry and mineralogy of an Indonesian coal during combustion: part 1 drop-tube observations, *Fuel Processing Technology*, Vol. 89, pp. 595-607.
7. F. E. Huggins, (2002), Overview of Analytical Methods for Inorganic Constituents in Coal, *International Journal of Coal Geology*, Vol. 50, pp. 169-214.
8. K. Prakash and A. Sridharan, (2009), Beneficial Properties of Coal Ashes and Effective Solid Waste Management, Practice Periodical of Hazardous, *Toxic and Radioactive Waste Management*, Vol. 13, pp. 239-248.
9. L. C. Ram and R. E. Masto, (2010), An Appraisal of the Potential Use of Fly Ash for Reclaiming Coal Mine Spoil, *Journal of Environmental Management*, Vol. 91, pp. 603-617.
10. F. S. Depoi, D. Pozebon, and W. D. Kalkreuth, (2008), Chemical Characterization of Feed Coals and Combustion-By-Products from Brazilian Power Plants, *International Journal of Coal Geology*, Vol. 76, pp. 227-236.
11. J. Levandowski and W. Kalkreuth, (2009), Chemical and Petro graphical Characterization of Feed Coal, Fly Ash and Bottom Ash from the Figueira Power Plant, Paraná, Brazil, *International Journal of Coal Geology*, Vol. 77, pp. 269-281.
12. S. Dai, L. Zhao, S. Peng (2010), Abundances and Distribution of Minerals and Elements in High-Alumina Coal Fly Ash from the Jungar Power Plant, Inner Mongolia, China, *International Journal of Coal Geology*, Vol. 81, pp. 320-332.
13. M. Ahmaruzzaman, (2010), A Review on the Utilization of Fly Ash, *Progress in Energy and Combustion Science*, Vol. 36, pp. 327-363.
14. M. K. Panigrahi1, P. K. Rana, R. I. Ganguly, R. R. Dash, (2016), Structural Transformation of Cured Pond Ash Geopolymers: A Novel Construction Material, In. *20th National Conference on Nonferrous Minerals and Metals*, 8-9th July 2016; Eds. Rakesh Kumar, K.K.Sahu & Abhilash, pp.132-137.
15. Y. S. Liu, L. L. Ma, Y. S. Liu, and G. X. Kong, (2006). Investigation of Novel Incineration Technology for Hospital Waste, *Environmental Science and Technology*, Vol. 40, pp.6411-6417.

16. Y. S. Liu, (2005), A Novel Incineration Technology Integrated with Drying, Pyrolysis, Gasification and Combustion of MSW and Ashes Vitrification, *Environmental Science and Technology*, Vol. 39, 3855.
17. E. P. A. Chinese, (1996), China Identification Standard for Hazardous Waste Toxicity (GB5085.3-1996), Chinese Environmental Science Press, Beijing.
18. S. B. Kondawar, A. D. Dahegaonkar, V. A. Tabhane, and D. V. Nandanwar, (2014), Thermal and Frequency Dependence Dielectric Properties of Conducting Polymer/Fly Ash Composites, *Advanced Materials Letters*, Vol. 5, 360.
19. A. Pattnaik, S. K. Bhuyan, S. K. Samal, A. Behera, and S. C. Mishra, (2016), Dielectric Properties of Epoxy Resin Fly Ash Composite, *IOP Conference Series: Materials Science and Engineering*, Vol. 115, 1.
20. M. Mishra, A. P. Singh, and S. K. Dhawan, (2013), Utilization of Fly Ash-A Waste By-product of Coal for Shielding Application, *Journal of Environmental Nanotechnology*, Vol. 2, 74.
21. J. P. Dhal, and S. C. Mishra, (2013), Investigation of Dielectric Properties of a Novel Hybrid Polymer Composite using Industrial and Bio-waste, *Journal of Polymer Composites*, Vol. 1, 22.
22. K. Singh, T. Quazia, S. Upadhyaya, and P. Sakharkarb, (2005), Development of Low Permittivity Material using Fly Ash, *Indian Journal of Engineering & Materials Sciences*, Vol. 12, 345.
23. D. E. Yıldız, and I. Dökme, (2011), Frequency and Gate Voltage Effects on the Dielectric Properties and Electrical Conductivity of Al/SiO$_2$/p-Si Metal-Insulator-Semiconductor Schottky Diodes, *Journal of Applied Physics*, Vol. 110, 014507.
24. J. C. Egbai, (2013), Kaolin Quantification in Ukwu-nzu and Ubulu-uku using Electrical Resistivity Method, *International Journal of Recent Research and Applied Studies*, Vol. 14, 10 pp.
25. O. A. Babalola, and T. Akomolafe, (2008), Effects of Kaolin Particle Size and Annealing Temperature on the Resistivity of Zinc-Kaolin Composite Resistors, *Journal Applied Science Environmental Management*, Vol. 12, 113.
26. D. Goski and M. Lambert, (2019), Engineering Resilience with Precast Monolithic Refractory Articles, *International Journal of Ceramic Engineering and Science*, Vol. 1, pp. 169-177.
27. L. Li and Y. Li, (2017), Development and Trend of Ceramic Cutting Tools from the Perspective of Mechanical Processing, *IOP Conf. Series: Earth and Environmental Science*, Vol. 94, pp.1-6.
28. M. Zhu, R. Ji, Z. Li, H. Wang, L. L. Liu, and Z. Zhang, (2016), Preparation of Glass Ceramic Foams for Thermal Insulation Applications from Coal Fly Ash and Waste Glass, *Construction and Building Materials*, Vol. 112, pp. 398-405.
29. H. E. Exner and E. Arzt, (1996), Sintering Processes in Physical Metallurgy, Vol. 3, Ed. R. W. Cahn and P. Haasen. 4th Ed. Elsevier Science, Amsterdam, pp. 2628-2662.

30. W. V. Siemens, (1966), Inventor and Entrepreneur: Recollections of Werner von Siemens, London, England, 1966.
31. T. Tanaka and T. Imai, 2013, Advances in nanodielectric materials over the past 50 years, *IEEE Electrical Insulation Magazine*, Vol. 1, pp. 10-23.
32. T. Tanaka, 2005, Dielectric nanocomposites with insulating properties, *IEEE Transactions on Dielectrics and Electrical Insulation*, Vol. 12, pp. 914-928.
33. E. C. Nzenwa and A. D. Adebayo, (2019), Analysis of Insulators for Distribution and Transmission Networks, *American Journal of Engineering Research (AJER)*, Vol.8, pp.138-145.
34. C. L. Goldsmith, A. Malczcwski, J. J. Yao, S. Chen, J. Ehmk, and D. H. Hinzel, (1999), RFMEMS-Based Tunable Filters, *International Journal of RF and Microwave Computer-Aided Engineering*, Vol. 9, 362.
35. G. Subramanyam, F. V. Keuls and F. A. Miranda, (1998), Novel K-band Tunable Microstrip v-band Pass Filter using Thin Film HTS/Ferroelectric/Dielectric Multilayer Configuration, *IEEE Microwave Guided Wave Letter*, Vol. 8, 78.
36. A. Tombak, J. P. Maria, F. T. Ayguavives, Z. Jin, G. T. Stauf, A.I. Kingon, and A. Mortazawi, (2003), Voltage-Controlled RF Filters Employing Thin Film Barium-Strontium Titanate Tunable Capacitors, *IEEE Transection Microwave Theory and Technology*, Vol. 51, 462.
37. M. K. Panigrahi, R. R. Dash, R. I. Ganguly, (2018), Development of Novel Constructional Material From Industrial Solid Waste as Geopolymer for Future Engineers, *IOP Conference Series: Materials Science and Engineering*, Vol. 410, pp.1-12.
38. M. K. Panigrahi, P. Kumar, B. Barik, D. Behera, S. K. Mohapatra, and H. Jha, (2016), Frequency Dependency of Developed Dielectric Material from Fly Ash: An Industrial Waste, In. *20th National Conference on Nonferrous Minerals and Metals*, 8-9th July 2016; Eds. Rakesh Kumar, K.K.Sahu & Abhilash, pp.143-150.
39. T. R. Naik, R. Kumar, B. W. Ramme, and R. N. Kraus, (2010), Effect of High-Carbon Fly Ash on the Electrical Resistivity of Fly Ash Concrete Containing Carbon Fibers, Second International Conference on Sustainable Construction Materials and Technologies June 28th, Università Politecnica delle Marche, Ancona, Italy.
40. Y. Liu, F. Zeng, B. Sun, P. Jia, and I. T. Graham, (2019), Structural Characterizations of Aluminosilicates in Two Types of Fly Ash Samples from Shanxi Province, North China, *Minerals*, Vol. 9, pp. 358-16.
41. P. Rożek, M. Król, and W. Mozgawa, (2018), Spectroscopic studies of fy ash-based geopolymers, *Spectrochim Acta - Part A Mol Biomol Spectrosc* Vol. 198, pp. 283-289, https://doi.org/10.1016/j.saa.2018.03.034.
42. S. Kumar, F. Kristály, and G. Mucsi, (2015), Geopolymerisation behaviour of size fractioned fly ash, *Adv Powder Technol*, Vol. 26, pp. 24-30, https://doi.org/10.1016/j.apt.2014.09.001.

43. Z. Kledyński, A. Machowska, B. Pacewska & I. Wilińska, (2017), Investigation of hydration products of fly ash-slag pastes, *Journal of Thermal Analysis and Calorimetry*, Vol. 130, pp. 351-363, https://doi.org/10.1007/s10973-017-6233-4.
44. D. Bondar and R. Vinai, (2022), Chemical and Microstructural Properties of Fly Ash and Fly Ash/Slag Activated by Waste Glass-Derived Sodium Silicate, *Crystals*, Vol. 12, pp. 913-12, https:// doi.org/10.3390/cryst12070913.
45. R. Roy, D. Das, and P. K. Rout, (2022), A Review of Advanced Mullite Ceramics, *Engineered Science*, Vol. 18, pp. 20-30.

7

High Resistance Sintered Pond Ash/Kaolin (PA/CC) Ceramics

M. K. Panigrahi[1]*, R.I. Ganguly[2] and R.R. Dash[3]

[1]Department of Materials Science, Maharaja Sriram Chandra Bhanja Deo University, Balasore, Odisha, India
[2]Department of Metallurgical Engineering, National Institute of Technology, Raurkela, Odisha, India
[3]CSIR-National Metallurgical Laboratory, Jamshedpur, Jharkhand, India

Abstract

In this work, pond ash is mixed with different weight fraction of kaolin (i.e., 10%/20%/30%/40%/50%). Mixture are compacted and sintered at three different temperatures (i.e., 1000 °C, 1100 °C, and 1200 °C). Chemical composition of raw pond ash and Kaolin (CC) are cited. Sintered composite are characterized by X-ray diffraction, FT infrared spectroscope, FESEM with EDS, and resistivity (room temperature). Structure, chemical group, and surface topography are examined. The best resistivity value has been observed for pondash (60%)/kaolin (40%) composite sintered at 1200 °C (1 h). The highest resistivity value is found to be 141.486345 Ω.cm. Mullite phase has been identified for sintered materials by XRD analyses. This phase enhances resistivity of sintered samples. Presences of elements of sintered materials are assessed by EDX analyses. Different chemical groups occurring in sintered materials have been shown through FTIR analyses.

Keywords: Industrial wastes, pond ash, Kaolin, XRD, mullite, morphology, resistivity and insulator

7.1 Introduction

Ceramics are a potential candidate for many technological applications. This is because of water absorption and open porosity behaviour of these

---

Corresponding author: muktikanta2@gmail.com

materials, which approaches close to zero [1]. Advantage of ceramic is that it retains its physico-mechanical properties i.e., hardness, abrasion resistance and bending strength. These characteristics are noticeable. Compositions of ceramics are related to mineralogical composition of raw materials [2].

Clays are abundantly available in nature [3]. They have many properties to justify their potential candidature in many technological applications (*i.e.* Civil construction and oil wells, chemical, white ceramic, food, drugs and cosmetics, filler for polymers, etc). Therefore, clays are used as minor component for preparing composites [4].

Kaolin is an important raw material used for preparation of ceramics. It is a low cost aluminosilicate mineral. It primarily occurs as clay sized particles with high aspect ratios. Hence, it may play very dynamic role to regulate degree of crystallinity, concentration of impurities, particles size distribution, etc. [5].

In spite of its economical and geological advantage, its spectroscopic account is not well recognized. Technological surveys indicate its availability with current market price [6]. Clay minerals such as kaolinite, bentonite, smecttite, diatomite and fullers occur naturally. It can substitute for activated carbon due to its availability, low cost, and good sorption properties [7]. Clays are layered structured minerals. Layers consist of different structural units. They are TO_4 tetrahedra (T = Si^{4+}, Al^{3+}, etc.) and MO_6 octahedra (M = Al^{3+}, Fe^{3+}, etc.). Layers belong in 1:1 clay minerals family and hold together by hydrogen bonds [8]. Structural units of clay consist of silica tetrahedral and octahedral units. These units are formed by octahedrally coordinated cations and oxygens or hydroxyls octahedra [8]. Natural clay contains impurities like- quartz, calcite, feldspar, mica organic matter, hydrated iron oxide, ferrous carbonate and pyrite with clay minerals. Physical properties of clay particles depend on size, shape, cation exchange capacity and type of impurities present in clay. Particle sizes vary from micrometer to nanometer range.

Thermal power plants produce large amount of waste [9]. Coal ashes are generated as wastes during combustion of coal [9, 10]. Physico-chemical properties of ash depend on source of coal and burning process of coal [11]. Occurrences of elements in large quantities in ashes are silicon, calcium, aluminum, iron, magnesium, sulfur, carbon. Elements present in small quantities are called trace elements [12]. Coal ash is categorized by three types *viz.* bottom ash, fly ash, and pond ash [13].

Bottom ash possesses threat to our environment. Thermal power station is experiencing disposal problem. Therefore, disposal of ash has become thrust area for engineering research. At present, ash is used as an additive to many manufacturing engineering applications such as manufacture of cement and concrete, conventional ceramics, and vitroceramics, etc [14]. At present, attempts are made to utilize waste for manufacturing of new products. To make this attempt feasible there are requirements of large mass flow and high temperature.

Thermal power plant wastes contain different elements (*i.e.*, Cu, Pb, Cd, Ag, Mo, Fi, Ti, Na, Mn, S, P, Zn, and Cl). These elements are present in their oxide form. Among these elements, some have insulating properties and others are semiconducting in nature [15, 16]. Therefore there is a possibility to use this materials for making electronic materials (*i.e.*, insulators and semiconducting).

Dielectric properties and DC resistivity of industrial waste based materials have been studied at higher temperature range (300-500 °C) [17].

Thermal power plant ash mainly comprises of crystalline phases and glassy phase [18]. Work is going on to improve dielectric and electrical insulating properties of industrial wastes [19–22].

DC resistivity, dielectric constant, and dielectric loss are main considerations for fabricating insulators [23].

Ceramics are used for different purposes *viz.*, refractories [24], cutting tools [25], thermal insulations [26], etc. Ceramic preparation requires higher temperature and pressure. Performances of ceramics are assessed by the presence of crystalline phase(s). Sintering process/temperature play an important role to achieve mullite phases. Mullite phase is formed between 1150 to 1300 °C [27]. Presence of mullite phase enhances insulating properties. Control of microstructure depends on composition of ash and sintering temperature. Optimization of process parameters for obtaining maximum amount mullite phase will enhance insulating properties.

At first, ceramic materials are used as electrical insulators in 1850 during the construction of electrical air lines [28]. Ceramics have number of distinctive properties (*i.e.*, mechanical strength, high dielectric strength, and corrosion resistance) [29, 30]. In present scenarios, low cost insulator has large demand in electrical engineering sectors. Current demands motivate researchers to develop insulators to meet present requirements [31]. Insulators have great importance for day-to-day life. They make live easier, harmless and shock free [32–34].

In the current work, sintered pond ash/kaolin composite is prepared through a sintering process (at 1200 °C) via solid state route. Sintered pond ash/kaolin composite is analyzed using FESEM with EDS. Phases of sintered composite are investigated by X-ray diffraction technique. FTIR test is done to identify the presence of chemical groups in the materials. Electrical resistivity value are also calculated.

7.2 Experimental Details

7.2.1 Materials and Chemicals

Pond ash is collected from NALCO, Damanjodi, Odisha, India. Kaolin, dextrin $(C_6H_{10}O_5)_n . xH_2O)$ are obtained from Loba chemicals. During composite preparation, 6% water is used. Compositions of as-processed pond ash, Kaolin are indicated in Table 7.1. Optical images of pond ash, kaolin, and dextrin are shown in Figure 7.1.

7.2.2 PA/Kaolin Composite Preparation

Preparation of pond ash-based sintered composite *viz.*, PA/Kaolin are achievable through solid state route. Following steps are adopted to obtain PA based composite [36, 37].

Step-1 preparation of processed pond ash
Collected pond ash is grinded in a ball mill (for 5 h). It is sieved to 240 meshes. Sieved pond ash is dried in a heating oven at 120 °C (2 h). Aim of heating sieved pond ash is to remove occulded moisture. This is termed as processed pond ash.

Step-2 Preparation of green pellet
Required amount of processed pond ash is placed in a mortar with pastel. Requisite amount of kaolin is added to the mortar and pastel containing pond ash. It is grinded properly. Small amount of dextrin (i.e., 0.05 %) and a few centimeter cubes (cc) of water are added to grinded mixture to make the mixtures pasty. Such paste samples are known as green samples. Samples are ready for preparing green pellets.

The assembled pelletizer are properly cleaned, washed, and dried before use. Desired pellets are made using pelletizer and universal testing machine (UTM). Appropriate amounts of samples are kept in the pelletizer. During the compaction, pressure (10 MPa) is applied to the sample containing pelletizer. It is left for five minutes for proper compaction. Then, pelletizer is removed from moulding system. Pellets are collected from pelletizer by demoulding the assembly. Dimensions (Thicknesses and diameter) of pellet(s) are measured. Such pellets are ready for sintering.

Step-3 Preparation of sintered pond ash based materials
Green pellet samples are kept in muffle furnace. They are bisquetted at 900 °C (for 1 h) to remove water from green pellets and binder. This process is

Table 7.1 Constituents with percentage (%) of processed pond ash and Kaolin.

S. ID	Compositions (%)												Ref.
	Fe_2O_3	Al_2O_3	SiO_2	P_2O_5	Cr_2O_3	ZnO	MgO	C	CaO	TiO_2	MnO	LOI	
PA	3.85	28.3	62.8	0.32	0.04	0.027	0.049	1.15	0.7	1.84	0.03	0.5	[35]
CC	---	37.66	44.80	1.54	---	---	T	---	0.50	0.60	---	14.33	[36]

Note: S.ID = Sample ID, PA = Pond ash, CC = Kaolin.

120 High Electrical Resistance Ceramics

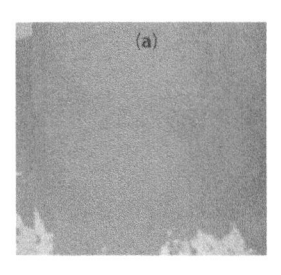

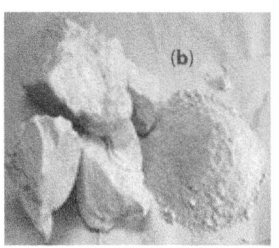

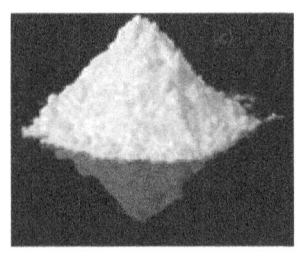

Figure 7.1 Optical image of pond ash (a), Kaolin (b) and Dextrin (c).

called bisquetting. Following bisquetted, pellet(s) are sintered at 1000 °C (for 2 h). Dimensions (*i.e,* thickness and diameter) of sintered pellet are found to be 0.5 cm and 3 cm, respectively. The process is shown by a flow chart (Figure 7.2).

Sintered pellets are ready for using different tests. Similar procedure is adopted for other sets of sintered products (Vide-Table 7.2).

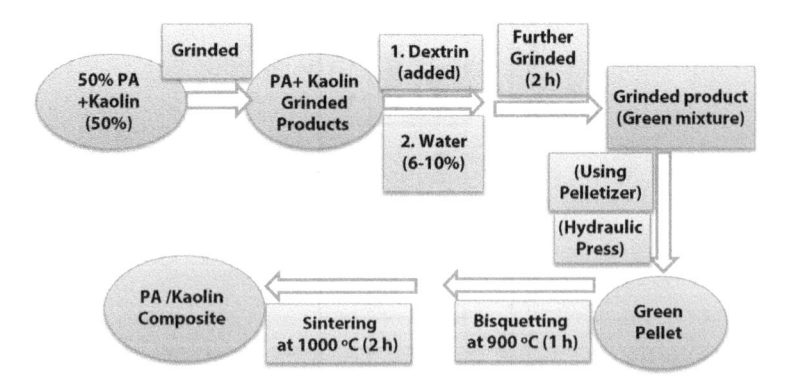

Figure 7.2 Flow chart for preparation of sintered pond ash/kaolin based composite.

Table 7.2 Compositions (%) of pond ash/CC composite materials preparation.

S. no.	Individual composition name	
	Pond ash	Kaolin (CC)
1	60	40
2	70	30
3	80	20
4	90	10

In order to find out effect of sintering temperature on resistivity value, samples are sintered at different sintering temperatures i.e., 1000 °C, 1100 °C and 1200 °C.

7.2.3 Test Methods

X-ray diffraction (XRD) analyses are carried out to identify different phases in sintered materials. For this, Phillips PW-1710 advanced wide angle X-ray diffractometer and Phillips PW-1729 X-ray generator are used. Cu target is chosen for CuKα radiation (wavelength (λ) is 0.154 nm). XRD machine is operated at 40 kV and 20 mA. Pellet samples are kept on a sample holder made of quartz. Measurement is carried out at room temperature. Samples are scanned at diffraction angles between 20° to 70° with a scanning rate of 2°/min.

Optical image is taken during hardness measurement by nano indentation (Anton Paar-NHT, Switzerland). Pellet sample of pond ash/Kaolin (60:40) composite is fixed on the moving sample holder. Morphological studies are performed with field emission scanning electron microscope (Carl Zeiss Supra 40). All sintered samples are Gold coated using sputtering technique. During the test, an instrument parameter *i.e.,* operating voltage is maintained at 30 kV. EDS data are also recorded during scanning of samples.

TEM (5022/22, TecnaiG220 S-Twin, Czech Republic) technique is used to investigate images and SAD pattern of raw pond ash (PA) powder and pond ash/kaolin (60:40) composite. TEM specimen (raw PA) is prepared by microtone technique. Also, TEM specimen of pond ash/k-Feldspar (60:40) composite is prepared by microtone technique (LEICA Microsystem, GmBH, A-1170). Then, specimens are transferred to Cu TEM grids.

FTIR spectra of pond ash (PA), kaolin, and sintered PA/kaolin composite are noted. Thermo Nicolet Nexus 870 spectrophotometer (range 400-4000 cm^{-1}) has performed this test. Settings parameters (*i.e.,* 50 scans at 4 cm^{-1} resolution and Absorbance measurement mode) are kept constant. For this test, pellet sample is made with sintered material mixed with KBr. Before the test, background spectrum is taken. Following which, pellet sample is placed in a FTIR sample holder and data are collected.

For electrical resistance measurement, sintered pellet samples are used. Resistivity is estimated using following relation [36, 37];

$$\rho = \frac{R \times A}{l} \quad (7.1)$$

Where, A is area of electrode (m^2), R is resistance of prepared samples (MΩ), l is thickness of sample, ρ is resistivity of the samples in Ω.cm. For measurement of resistivity, multimeter is used.

7.3 Results and Discussion

Resistivity values of sintered pond ash and sintered pond Ash/kaolin composite (at 1000 °C) having different proportions are shown in Table 7.3. It is observed from Table 7.3 that pond ash/kaolin composite (60:40) has fetched maximum resistivity value *i.e.*, 24.59403 × 10^9 ohm.cm, whereas pond ash sintered at 1000 °C shows lowest resistivity value and is to be 4.947075 × 10^6 ohm.cm.

Table 7.4 show resistivity values of pond ash and pond Ash/kaolin with different proportions. Sintering temperature has been increased to 1100 °C. It is seemed that resistivity value of PA/Kaolin (60:40) ratio has increased.

Therefore, with increase of sintering temperature, resistivity value has further improved in comparison to resistivity value measured for same composition of composite sintered at 1000C.

Table 7.3 Resistance value of pond ash/Kaolin based samples [radius=1.5 cm and thickness=0.5 cm].

S. no.	Sample ID	Sintering temperature (°C)	Resistance	Resistivity value (Ω.cm)
1	Pond ash	1000	0.35 MΩ	4.947075 × 10^6
2	Pond ash (50%)/ Kaolin (50%)	1000	0.15 GΩ	2.120175 × 10^9
3	Pond ash (60%)/ Kaolin (40%)	1000	1.74 GΩ	24.59403 × 10^9
4	Pond ash (70%)/ Kaolin (30%)	1000	1.64 GΩ	23.18058 × 10^9
5	Pond ash (80%)/ Kaolin	1000	0.92 GΩ	13.00374 × 10^9
6	Pond ash (90%)/ Kaolin (10%)	1000	0.07 GΩ	0.989415 × 10^9

Table 7.4 Resistance value of pond ash/Kaolin based samples [radius=1.5 cm and thickness=0.5 cm].

S. no.	Sample ID	Sintering temperature (°C)	Resistance	Resistivity value (Ω.cm)
1	Pond ash	1100	0.95 MΩ	13.427775×10^6
2	Pond ash (50%)/ Kaolin (50%)	1100	0.58 GΩ	8.19801×10^9
3	Pond ash (60%)/ Kaolin (40%)	1100	2.54 GΩ	35.90163×10^9
4	Pond ash (70%)/ Kaolin (30%)	1100	1.41 GΩ	19.929645×10^9
5	Pond ash (80%)/ Kaolin	1100	0.83 GΩ	11.731635×10^9
6	Pond ash (90%)/ Kaolin (10%)	1100	0.51 GΩ	7.208595×10^9

Table 7.5 shows resistivity values of different of pond ash/Kaolin Composite sintered with different proportions at 1200 °C. It is observed that pond ash and kaolin in 60:40 has achieved the maximum resistivity value at a sintering temperature 1200 °C. Thus, there is around 4.5-fold enhancement (10.16 GΩ) of resistivity value for the composite (60:40 compositions) sintered at 1200 °C. Therefore, resistivity value is further improved with increasing sintering temperature. This is due to formation of mullite phase in higher quantities at 1200 °C sintering temperatures [38].

From Table 7.3, Table 7.4, and Table 7.5, it is concluded that 1200 °C sintered pond ash with 60 (pond ash):40 (kaolin) proportion composite is shown highest resistivity value i.e., 141.486345×10^9.Ohm.cm.

Figure 7.3 shows superimposed XRD pattern of pond ash and pond ash/kaolin composite (three different compositions) sintered at 1200 °C. In XRD pattern (Figure 7.3a), Different phases identified are alumina, silica, and mullite. The highest intensity peak identified for pond ash is silica (Figure 7.3a). A few other peaks are matched (JCPS data file) with mullite and alumina phases. Table 7.6 shows crystal system and crystallographic parameters for identified phases. Figures 7.3b, 7.3c, and 7.3d show XRD pattern of composites of pond ash and kaolin with three different compositions. All are sintered at 1200 °C. Figures 7.3b shows appearance of a peak

124 High Electrical Resistance Ceramics

Table 7.5 Resistance value of pond ash/Kaolin based samples [radius=1.5 cm and thickness=0.5 cm].

S. no.	Sample ID	Processing temperature (°C)	Resistance	Resistivity (Ω.cm)
1	Pond ash	1200	0.58 GΩ	8.19801×10^9
2	Pond ash (50%)/ Kaolin (50%)	1200	1.01 GΩ	14.275845×10^6
3	Pond ash (60%)/ Kaolin (40%)	1200	10.16 GΩ	141.486345×10^9
4	Pond ash (70%)/ Kaolin (30%)	1200	5.3 GΩ	74.91285×10^9
5	Pond ash (80%)/ Kaolin	1200	2.61 GΩ	36.891045×10^9
6	Pond ash (90%)/ Kaolin (10%)	1200	1.19 GΩ	16.820055×10^9

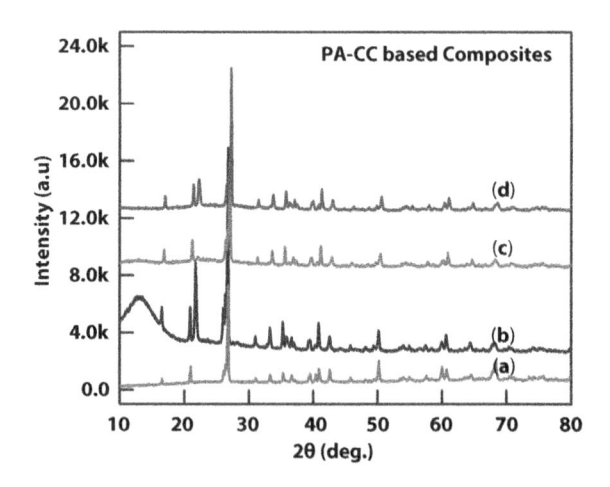

Figure 7.3 XRD pattern of pond ash (a), PA(60%)/CC(40%) composite (b), PA(70%)/CC(30%) composite (c), PA(80%)/CC(20%) composite (d) [Where CC~kaolin].

(13°) with broadens line profile. Broaden line profile is due to non-ideal diffraction presumably occuring for small crystal size materials. Some peaks have matched with silica phases. Other peaks occur due to interaction between oxides present in pond ash and kaolin. Also, XRD pattern other compositions do not show occurrence of any non-ideal peak (13°) if compared XRD pattern of Figure 7.3b. Other Small x-ray peaks are occurring in Figure 7.3c, 7.3d are similar that of peaks occurring in Figure 7.3b.

Strong line for silica phase has occurred for all sintered composite samples. Occurrences of many other peaks in sintered products are due to interaction between phases present in composite compositions. XRD pattern of three composite materials are shown Figure 7.3b, 7.3c, 7.3d. On comparison, it is evident that mullite phases at pond ash (60%): Kaolin (40%) are stronger having higher and prominent peaks. At other composition, intensity lines have faded with smaller peak height. It is conclusive that both glassy phase and higher proportion of mullite phase have occurred due to formation of mullite phase. Mullite phase belongs to the orthorhombic crystal system. Crystallographic parameters are presented in Table 7.6.

Optical image of pond ash/Kaolin (60:40) composite (sintered at 1200 °C) is displayed in Figure 7.4. It shows irregular shaped particles of different sizes. Owing to poor resolution, detail of image is not clearly revealed.

Figure 7.5a and 7.5b show FESEM images of pond ash and kaolin, respectively. Figure 7.5a shows ellipsoid particles having different aspect ratios. Particle sizes vary widely. The highest major axis and minor axis are measured to be 16 micrometer to 8 micrometers. For smallest one, major axis is one micron and minor axis is half micron. The aspect ratio is around 2:1.

Figure 7.5b shows micrograph of kaolin. It is an irregular shaped structure. It consists of different oxide compounds. There is distinct difference between microstructure of pond ash and kaolin.

On sintering of pond ash and kaolin in the ratio of 60:40 (1200 °C), there is marked difference between sintered and unsintered products. Figure 7.5c shows sintered product at a lower magnification. Owing to sintering, there is reaction between phases present in raw materials. Microstructure becomes dense and coherent. In order to understand morphology of phases in sintered product (Figure 7.5c), it is examined at a very high magnification. It can be noticed that there are crystalline and glassy phases coexistence in the microstructure.

EDS analysis of sintered pond ash/kaolin (60:40) composite are performed simultaneously with TEM studies. EDS profile (Figure 7.6) identifies presence of elements (i.e., O, Al, Si, Ti, Fe, K, etc). Elements like O, Al, and Si are prominent elements, however Ti, Na, K, etc are found in traces. EDS results have supported to XRD results.

Table 7.6 Structural information of pond ash/Kaolin based composite.

Crystal ID	Mineral name	Crystal system	Space group	Space group number	Cell dimension	Crystal angle	Calculated density (g/cm³)
As-received PA	Quartz SiO_2	Hexagonal	P3221	154	a=b≠c	α=β=90, γ=120	2.65
PA 1200 °C	Mullite $Al_2(Al_{2.5}Si_{1.5})O_{9.75}$	Orthorhombic	Pbam	55	a≠b≠c	α=β=γ=90	3.17
PA 1200 °C	Quartz SiO_2	Cubic	Fd-3m	227	a=b=c	α=β=γ=90	2.21
PA+CC (60%+40%) 1200 °C	Mullite $3Al_2O_3 \cdot 2SiO_2$	Orthorhombic	Pbam	55	a≠b≠c	α=β=γ=90	3.16
PA+CC (60%+40%)	Quartz SiO_2	Hexagonal	P3121	152	a=b≠c	α=β=90, γ=120	2.66

Note: PA-Pond ash, CC-Kaolin, SiO_2-Silicon dioxide.

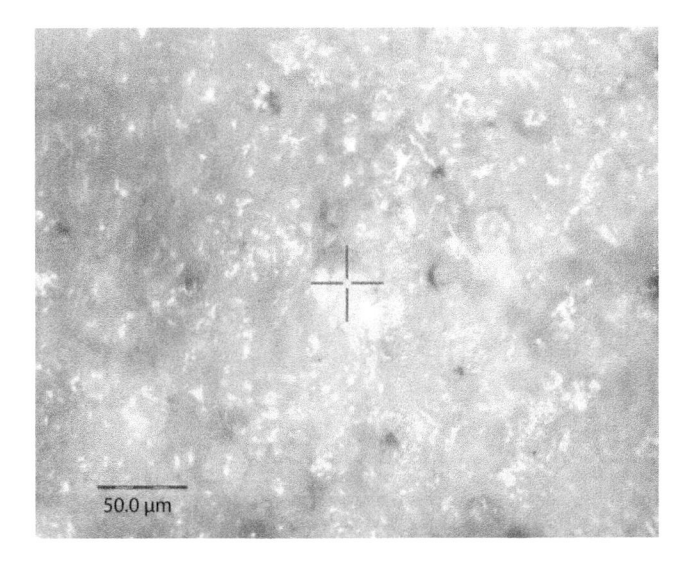

High Resistance Sintered Pond Ash/Kaolin (PA/CC) Ceramics 127

Figure 7.4 Optical image of pond ash/CC (60%-40%) based composite.

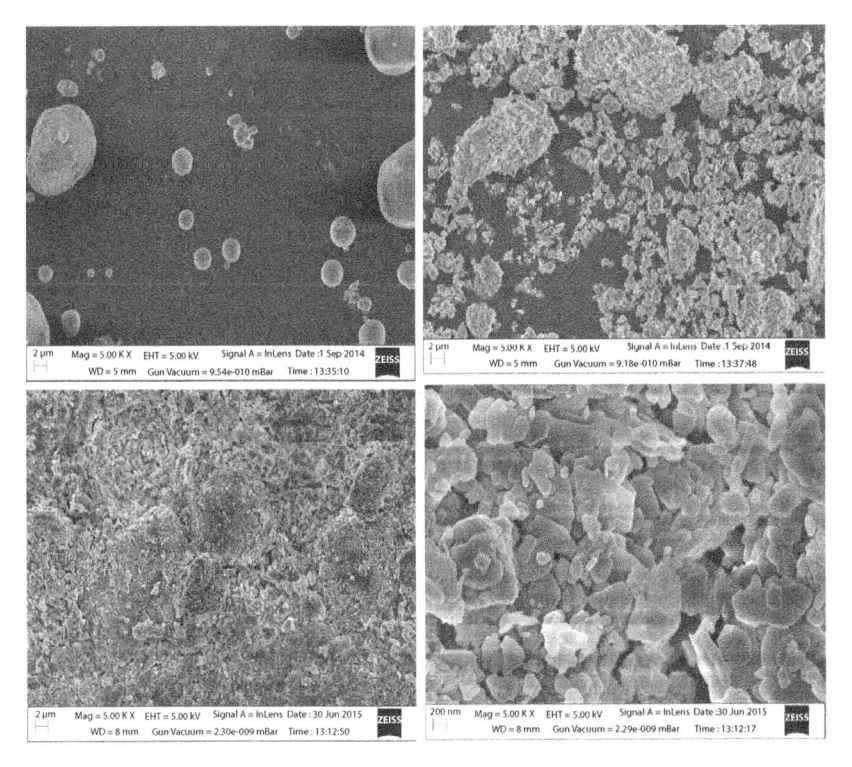

Figure 7.5 PA (a), Kaolin (b), PA/kaolin (lower magnification, c), PA/kaolin (higher magnification, d).

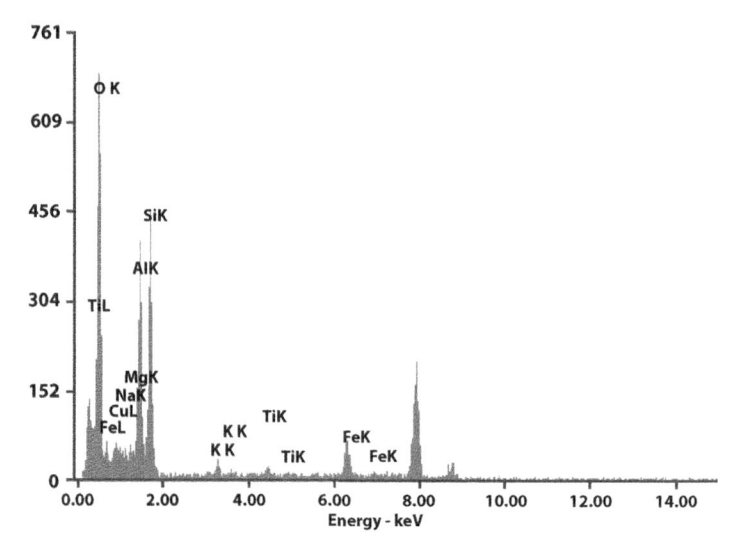

Figure 7.6 EDS spectrum of pond ash/Kaolin (60:40) composite.

Figure 7.7a and 7.7b indicate TEM micrographs of Pond Ash (A) and pond ash (60%)/kaolin (40%) sintered composite at 1200 °C. Figure 7.7a shows TEM micrograph of one of the embeded particles of pond ash. Embeded structure looks like ellipse having average major axis (135 nm) and average minor axis (65 nm). On sintering of pond ash and kaolin in the ratio 60:40 at a temperature of 1200 °C, there is interaction between

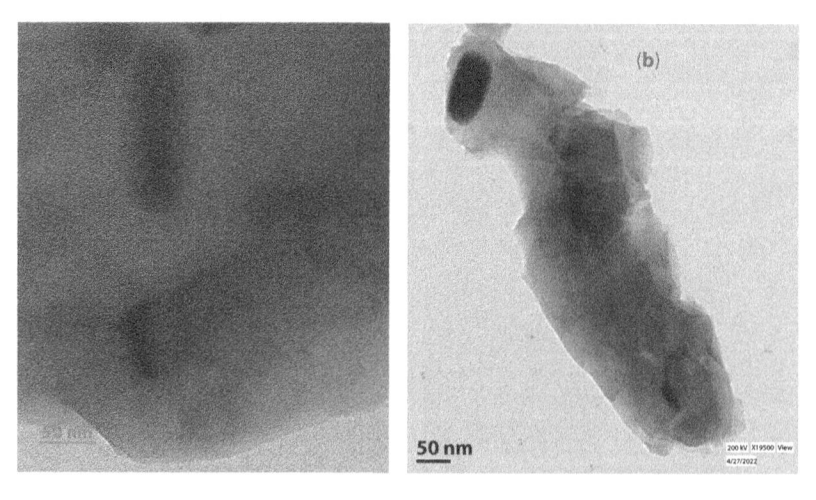

Figure 7.7 TEM images of pond ash (a) and pond ash (60%)/kaolin (40%) sintered at 1200 °C (b).

these materials. Magnified view of the composite can be observed in TEM photograph, which is shown in Figure 7.7b. Evidently, picture shows a fused mass formed due to sintering of mixture.

Corroseponding selected area diffraction pattern of pond ash and pond ash/kaolin composite (sintered at 1200 °C) are shown in Figure 7.8a and 7.8b. Both the figures show bright spots indicating crystalline phases present in two materials. There is more number of bright spots observed for pond ash materials in comparisons with sintered material. Microscopically, sintered materials contain more amount of glassy phase if compared with pond ash. This can be corroborated with XRD analyses, which is discussed earlier (Figure 7.3).

FTIR absorption bands of pond ash are studied [39]. Different absorption stretching frequency of pond ash powder corresponds to pond ash components. The stretching frequency at 600 cm^{-1} matches to Si-O-Al band [40]. Stretching frequencies are found at 1098 and 1608 cm^{-1} may be acrideted to Si-O-Si asymmetric band [41] and H-O-H bending band [42] respectively.

FTIR spectrum of kaolin is shown in Figure 7.9a. From Figure 7.9a, broad stretching bands found around 2919.50 cm^{-1}. The band arises due to the stretching vibrations of O-H bonds and H bending vibrations of H-O-H of interlayer adsorbed H_2O molecule [42]. C=O stretching band is indicated at 1449.40 cm^{-1} and confirms the presence of carbonate groups [40]. The main reason being the presences of chemisorbed CO_2 in pond ash/kaolin. Si-O and O-Si-O bands [43] are also found at 989.30 cm^{-1} and 534.50 cm^{-1}. It indicates the presence of silicate groups. Presence of

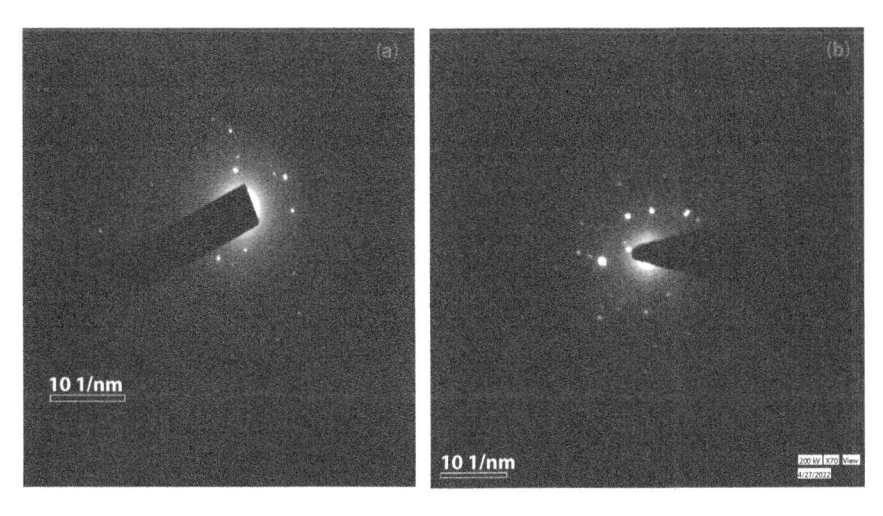

Figure 7.8 SAD profile of pond ash (a) and pond ash (60%)/kaolin (40%) sintered at 1200 °C (b).

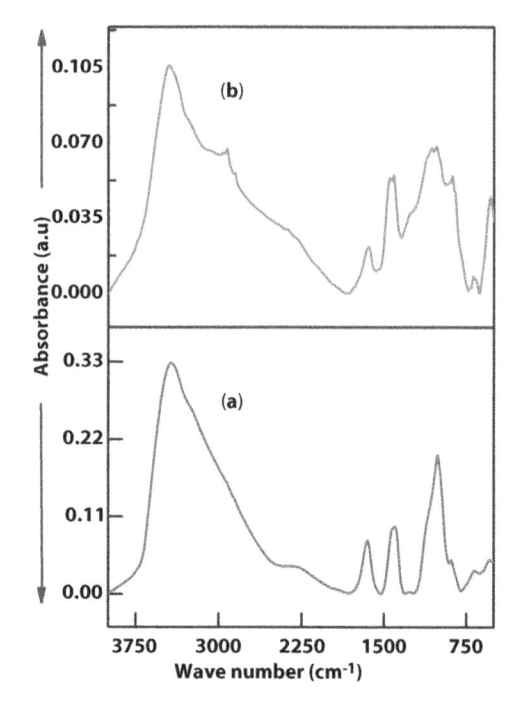

Figure 7.9 Sintered pond ash (a) and pond ash (60%)/kaolin (40%) composite (b) sintered at 1200 °C.

Al^{3+}-O^{2-} bonds [44] are also observed near 805.5 cm^{-1}. Fe-O stretching band [45] is also observed at 440 cm^{-1}.

The FTIR spectrum of pond ash (60%)/kaolin (40%) composite is shown in Figure 7.9b. Figure 7.9b indicates the presence of different absorption bands occurring at different wave numbers i.e., 3441, 2918, 1823, 1638, 1117 and 907 cm^{-1}. Bands are in agreement with stretching vibrations of O-H bonds (3441 cm^{-1} wavenumber) and H-O-H bending vibrations (1638 cm^{-1} wave number) of interlayer adsorbed H_2O molecule [42]. The hydroxyl-stretching band of water plays an important role and peak shift of the FTIR spectra is significant. Absorption band at 1117 cm^{-1} wave number is attributed to the Si-O band [43] and signifies the occurrence of silicate groups. Presence of $Al^{3+}$$O^{2-}$ absorption bands is also indicated near 907 cm^{-1} wavenumber [44].

7.4 Conclusions

Pond ash is mixed with different weight fraction of kaolin (10%, 20%, 30%, 40%, and 50%). Samples are sintered in muffle furnace in three temperatures. Microstructural transformation is observed in FESEM analyses. FESEM analyses of pond ash show elliptical shaped pond ash particles in different ratios. Aspect ratio i.e., ratio of major to minor axis is found to be approximately 2:1. Sintered pond ash mixed with kaolin shows matrix consisting of dense glassy phase with some crystalline phases. Procured kaolin shows flaky structure. EDS analyses show presence of elements like O, Si, Al, Mg, Ca, Ti, etc. Mullite and other phases are clearly indicated in the XRD pattern. Unit cell structure of mullite and quartz are orthorhombic and Tetragonal, respectively. Estimated density of orthorhombic (obtained from mullite) and tetragonal (obtained quartz phase) unit cells are 3.16 g/cm^3 and 2.26 g/cm^3, respectively. Noticeable variance in FTIR spectra is observed. PA/kaolin composite (at 1200 °C) has shown with highest electrical resistivity and is to be 141.486345 × 10^9 Ω.cm.

Acknowledgements

First author would like to thank Prof. Munesh Chandra Adhikary, PG Council Chairman, Fokir Mohan University, for his invaluable guidance, advice, and constant inspiration throughout the entire program. First author also thanks Mr. Mukteswar Mohapatra, Fokir Mohan University for their active support during preparation of manuscript. The authors convey their sincere thanks to GIET, University Gunupur, Rayagada, Odisha, India for providing Lab facilities to carry out the research work. Authors also wish to thank the CRF, IIT Kharagpur for providing testing facilities.

References

1. A. R. G. Azevedo, C. M. F. Vieira, W. M. Ferreira, K. C. P. Faria, L. G. Pedroti, B. C. Mendes, (2020), Potential Use of Ceramic Waste as Precursor in the Geopolymerization Reaction for the Production of Ceramic Roof Tiles, *Journal of Building Engineering*, Vol. 29, 101156, https://doi.org/10.1016/j.jobe.2019.101156.
2. Sk S. Hossain and P. K. Roy, (2020), Sustainable Ceramics Derived from Solid Wastes: A Review, *Journal of Asian Ceramic Societies*, Vol. 8, pp. 984-1009, https://doi.org/10.1080/21870764.2020.1815348.
3. F. Bergaya and G Lagaly, (2006), General Introduction: Clays, Clay Minerals and lay science, *Handbook of clay science*, Vol. 1, pp. 1-8.
4. C. N. Djangang, A. Elimbi, U. C. Melo, G. Lecomte-Nana, C. Nkoumbou, J. Soro, J. P. Bonnet, P. Blanchart, and D. Njopwouo, (2008), Sintering

of Clay-Chamote Ceramic Composites for Refractory Bricks, *Ceramics International*, Vol. 34, pp. 1207-1213.
5. S. K. Hubadillah, Z. Harun, Mohd H. D. Othman, and A. F. Ismail, and P. Gani, (2016), Effect of Kaolin Particle Size and Loading on the Characteristics of Kaolin Ceramic Support Prepared Via Phase Inversion Technique, *Journal of Asian Ceramic Societies*, Vol. 4.
6. J. (Theo) Kloprogge, Spectroscopic Methods in the Study of Kaolin Minerals and Their Modifications, Springer Mineralogy, pp.1-428, https://doi.org/10.1007/978-3-030-02373-7.
7. S. Mirmohamadsadeghi, T. Kaghazchi, M. Soleimani, and N. Asasian, (2012), An Efficient Method for Clay Modification and Its Application for Phenol Removal from Wastewater, *Applied Clay Science*, Vol. 59-60, pp. 8-12, https://doi.org/10.1016/j.clay.2012.02.016.
8. Blanca Bauluz Lázaro, (2015), Halloysite and Kaolinite: Two Clay Minerals with Geological and Technological Importance, *Review of Real Academia de Ciencias. Zaragoza.* Vol. 70, pp. 1-33.
9. Thermal Power Plants-Advanced Applications, Edited by Mohammad Rasul, April 17th, 2013, pp.1-188, IntechOpen.
10. http://www.groundtruthtrekking.org/Issues/AlaskaCoal/Coal-Ash-Combustion-Wastes.html
11. M. Farhad Howladar & Md. Raisul Islam, (2016), A study on Physico-Chemical Properties and Uses of Coal Ash of Barapukuria Coal Fired Thermal Power Plant, Dinajpur, for Environmental Sustainability, *Energy, Ecology and Environment*, Vol. 1, pp. 233-247.
12. A. Fuller, J. Maier, E. Karampinis, J. Kalivodova, P. Grammelis, E. Kakaras, and G. Scheffknecht, (2018), Fly Ash Formation and Characteristics from (co-)Combustion of an Herbaceous Biomass and a Greek Lignite (Low-Rank Coal) in a Pulverized Fuel Pilot-Scale Test Facility, *Energies*, Vol. 11, pp. 1581-38.
13. S. Vassilev and C. G. Vassileva, (2007), A New Approach for The Classification of Coal Fly Ashes based on Their Origin, Composition, Properties, And Behavior, *Fuel*, Vol. 86, pp. 1490-1512.
14. R. P. dos Santos, J. Martins, C. Gadelha, B. Cavada, A. V. Albertini, F. Arruda, M. Vasconcelos, E. Teixeira, F. Alves, J. L. Filho, and V. Freire, (2014), Coal Fly Ash Ceramics: Preparation, Characterization, and Use in the Hydrolysis of Sucrose, *Scientific World Journal*, Vol. 2014, pp.1-7.
15. O. Popov, A. Iatsyshyn, V. Kovach, V. Artemchuk, I. Kameneva, O. Radchenko, K. Nikolaiev, V. Stanytsina, A. Iatsyshyn, and Y. Romanenko, (2021), Effect of Power Plant Ash and Slag Disposal on the Environment and Population Health in Ukraine, *Journal of Health and Pollution*, Vol. 11, pp. 1-10.
16. M. S. Chavali and Maria P. Nikolova, (2019), Metal Oxide Nanoparticles and Their Applications in Nanotechnology, *SN Applied Sciences*, Vol. 1, pp.1-30, https://doi.org/10.1007/s42452-019-0592-3.

17. E. P. A. Chinese, (1996), China Identification Standard for Hazardous Waste Toxicity (GB5085.3-1996), Chinese Environmental Science Press, Beijing.
18. S. B. Kondawar, A. D. Dahegaonkar, V. A. Tabhane, and D. V. Nandanwar, (2014), Thermal and Frequency Dependence Dielectric Properties of Conducting Polymer/Fly Ash Composites, *Advanced Materials Letters*, Vol. 5, 360.
19. A. Pattnaik, S. K. Bhuyan, S. K. Samal, A. Behera, and S. C. Mishra, (2016), Dielectric Properties of Epoxy Resin Fly Ash Composite, *IOP Conference Series: Materials Science and Engineering*, Vol. 115, p 1.
20. M. Mishra, A. P. Singh, and S. K. Dhawan, (2013), Utilization of Fly Ash-A Waste By-product of Coal for Shielding Application, *Journal of Environmental Nanotechnology*, Vol. 2, 74.
21. J. P. Dhal, and S. C. Mishra, (2013), Investigation of Dielectric Properties of a Novel Hybrid Polymer Composite using Industrial and Bio-waste, *Journal of Polymer Composites*, Vol. 1, 22.
22. K. Singh, T. Quazia, S. Upadhyaya, and P. Sakharkarb, (2005), Development of Low Permittivity Material using Fly Ash, *Indian Journal of Engineering & Materials Sciences*, Vol. 12, 345.
23. D. E. Yıldız and I. Dokme, (2011), Frequency and Gate Voltage Effects on the Dielectric Properties and Electrical Conductivity of Al/SiO2/p-Si Metal-Insulator-Semiconductor Schottky Diodes, *Journal of Applied Physics*, Vol. 110, 014507.
24. J. C. Egbai, (2013), Kaolin Quantification in Ukwu-nzu and Ubulu-uku using Electrical Resistivity Method, *Conference Proceedings International Journal of Recent Research and Applied Studies*, Vol. 14.
25. O. A. Babalola and T. Akomolafe, (2008), Effects of Kaolin Particle Size and Annealing Temperature on the Resistivity of Zinc-Kaolin Composite Resistors, *Journal Applied Science Environmental Management*, Vol. 12, 113.
26. L. Li and Y. Li, (2017), Development and Trend of Ceramic Cutting Tools from the Perspective of Mechanical Processing, *IOP Conf. Series: Earth and Environmental Science*, Vol. 94, pp.1-6.
27. M. Zhu, R. Ji, Z. Li, H. Wang, L. L. Liu, and Z. Zhang, (2016), Preparation of Glass Ceramic Foams for Thermal Insulation Applications from Coal Fly Ash and Waste Glass, *Construction and Building Materials*, Vol. 112, pp. 398-405.
28. H. E. Exner and E. Arzt, (1996), Sintering Processes, In Physical Metallurgy, Vol. 3, Ed. R. W. Cahn and P. Haasen, 4th Ed. Elsevier Science, Amsterdam, pp. 2628-2662.
29. W. V. Siemens, Inventor and Entrepreneur: Recollections of Werner Von Siemens, London, England, 1966.
30. T. Tanaka and T. Imai, (2013), Advances in Nanodielectric Materials Over the Past 50 years, *IEEE Electrical Insulation Magazine*, Vol. 1, pp. 10-23.

31. T. Tanaka, (2005), Dielectric Nanocomposites with Insulating Properties, *IEEE Transactions on Dielectrics and Electrical Insulation*, Vol. 12, pp. 914-928.
32. E. C. Nzenwa and A. D. Adebayo, (2019), Analysis of Insulators for Distribution and Transmission Networks, *American Journal of Engineering Research* (AJER), Vol. 8, pp. 138-145.
33. C. L. Goldsmith, A. Malczcwski, J. J. Yao, S. Chen, J. Ehmk, and D. H. Hinzel, (1999), RFMEMS-Based Tunable Filters, *International Journal of Radio Frequency and Microwave Computer-Aided Engineering*, Vol. 9, pp. 362-xxx.
34. G. Subramanyam, F. V. Keuls, and F. A. Miranda, (1998), Novel K-band Tunable Microstrip v-band Pass Filter using Thin Film HTS/Ferroelectric/Dielectric Multilayer Configuration, *IEEE Microwave Guided Wave Letter*, Vol. 8, 78.
35. M. K. Panigrahi, R. R. Dash, R. I. Ganguly, (2018), Development of Novel Constructional Material From Industrial Solid Waste as Geopolymer for Future Engineers, *IOP Conference Series: Materials Science and Engineering*, Vol. 410, pp.1-12.
36. M. K. Panigrahi, (2021), Investigation of Structural, Morphological, Resistivity of Novel Electrical Insulator: Industrial Wastes, *Bulletin of Scientific Research*, Vol. 3, pp. 51-58.
37. M. K. Panigrahi, (2022), Investigation of Structures of Sintered Fly Ash Materials: Resources of Industrial Wastes, *Bulletin of Scientific Research*, Vol. 4, pp.1-10.
38. V. Viswabaskaran, F. D. Gnanam, and M. Balasubramanian, (2004), Mullite from Clay-Reactive Alumina for Insulating Substrate Application, *Applied Clay Science*, Vol. 25, pp. 29-35, https://doi.org/10.1016/j.clay.2003.08.001.
39. A. Jose, M. R. Nivitha, M. Krishnan, and R. G. Robinson, (2020), Characterization of cement stabilized pond ash using FTIR spectroscopy, *Construction and Building Materials*, Vol. 263, 120136.
40. Y. Liu, F. Zeng, B. Sun, P. Jia, and I. T. Graham, (2019), Structural Characterizations of Aluminosilicates in Two Types of Fly Ash Samples from Shanxi Province, North China, *Minerals*, Vol. 9, pp. 358-16 .
41. P. Rożek, M. Król, and W. Mozgawa, (2018), Spectroscopic studies of fly ash-based geopolymers, *Spectrochimica Acta Part A: Molecular and Biomolecular spectroscopy*, Vol. 198, pp. 283-289, https://doi.org/10.1016/j.saa.2018.03.034.
42. S. Kumar, F. Kristály, and G. Mucsi, (2015), Geopolymerisation Behaviour of Size Fractioned Fly Ash, *Advanced Powder Technology*, Vol. 26, pp. 24-30, https://doi.org/10.1016/j.apt.2014.09.001.
43. Z. Kledyński, A. Machowska, B. Pacewska & I. Wilińska, (2017), Investigation of Hydration Products of Fly Ash-Slag Pastes, *Journal of Thermal Analysis and Calorimetry*, Vol. 130, pp. 351-363, https://doi.org/10.1007/s10973-017-6233-4.

44. D. Bondar and R. Vinai, (2022), Chemical and Microstructural Properties of Fly Ash and Fly Ash/Slag Activated by Waste Glass-Derived Sodium Silicate, *Crystals*, Vol. 12, pp. 913-12, https:// doi.org/10.3390/cryst12070913.
45. M. Gotic and S. Music (2007), Mossbauer, FT-IR and FE SEM Investigation of Iron Oxides Precipitated from $FeSO_4$ Solutions, *Journal of Molecular Structure*, Vol. 834-836, pp. 445-45.

8

High Resistance Sintered Pond Ash/Pyrophyllite (PA/PY) Ceramics

M. K. Panigrahi[1]*, R.I. Ganguly[2] and R.R. Dash[3]

[1]Department of Materials Science, Maharaja Sriram Chandra Bhanja Deo University, Balasore, Odisha, India
[2]Department of Metallurgical Engineering, National Institute of Technology, Raurkela, Odisha, India
[3]CSIR-National Metallurgical Laboratory, Jamshedpur, Jharkhand, India

Abstract

In this investigation, pond ash is mixed with pyrophyllite in different proportions and is sintered at 1200 °C. Before sintering, mixed materials are bisquetted (at 900 °C for 2 h) and by sintered at 1200 °C for 1 h. Sintered materials are characterized by X-ray diffraction (XRD), Fourier transformation infra-red spectroscope (FTIR) and field emission scanning electron microscope (FESEM) instruments. X-ray diffraction has revealed mullite phase formation due to interaction between pond ash and pyrophyllite. The Mullite phase enhances insulating property. FESEM analyses have revealed microstructural features of sintered products. FTIR analyses have identified different chemical group of sintered material. Maximum electrical resistivity at room temperature has been observed as a combination of two materials of pond ash and pyrophyllite in the ratio 60:40. It is found to be 38.587185×10^9 ohm.cm.

Keywords: Industrial wastes, pond ash, pyrophyllite, XRD, phases, mullite, morphology insulator, electrical resistance

8.1 Introduction

Uses of ceramic materials are well-known. They are used for manufacturing of refractories [1], cutting tools [2], thermal insulations [3], etc. Processing

---

Corresponding author: muktikanta2@gmail.com

of ceramics needs high application of pressure and sintering temperature. Microstructural constituents of ceramic materials are dominated by crystalline phases. Presence of crystalline phase(s) enhances properties of ceramic materials. It is observed that ceramic materials containing mullite phase possess high electrical resistivity. Therefore, it is important to select raw materials which on sintering will yield maximum mullite phase by interaction of oxides in raw materials. For this requirement, sintering temperature is selected between 1150 to 1300 °C [4].

Ceramic materials were used first as insulators in the year 1850 while constructing electrical air lines [5]. Distinctive properties of ceramics are; mechanical strength, high dielectric properties, corrosion and abrasion resistance [6, 7]. In current scenarios, low cost insulators are desired for application in electrical engineering sectors. Current demand motivates researchers to develop insulators having best properties to meet present requirements [8]. With increasing in industrialization and requirement of power sectors, demands for insulators have increased to a large extent. Insulators make life easier, harmless and shock free [9–11].

With proper sintering of compacted ceramic mass, water absorption and porosity are achievable close to zero [12]. Such material will have high value of physical and mechanical properties [13] i.e., hardness, abrasion resistance, bend strength and chemical inertness. These achievable properties of sintered ceramics are due to mineralogical composition of raw materials i.e., pyrophyllite and pond ash. Pyrophyllite is a layered aluminum silicate $[Al_2Si_4O_{10}(OH)_2]$. It belongs to 2:1 structure. Its structure comprises of $Al(O, OH)_6$ octahedra sheet and tetrahedra sheet. Octahedra sheet is sandwiched between two SiO_4 tetrahedra sheet. Pyrophyllite is made up of SiO_2 (67%), Al_2O_3 (28%) and H_2O (5%) [14–16]. In pyrophyllite ore, impurities are quartz, feldspar, diaspore, dickite, chlorite and mica [17–19]. Pyrophyllite is potential candidate for manufacturing ceramic, plastic, rubber, pressure transmission medium, medical carrier and adsorption or photocatalyst. This is due to its good physico-chemical characteristics (i.e., low thermal and electrical conductivity, low expansion coefficient, low reversible thermal expansion and excellent reheating stability) [15, 18, 20–23]. Therefore, they have influenced on electrical properties [24–31].

Flotation behavior of minerals is influenced by surface zeta potential [27, 32–34]. Zeta potential represents electric potential at the shear plane between a particle and surrounding liquid when charged particle moves in an electric field [32–34]. It depends on pH over a wide range [35]. Thus, charge behaviour of pyrophyllite should be considered in both basal plane surface and edge plane surface. Excess permanent negative charge of basal surface is attributed to isomorphic substitution of Si^{4+} replaced by Al^{3+} and Fe^{3+} in tetrahedral layers and Al^{3+} by Mg^{2+} and Fe^{2+} in octahedral layer [14, 36].

Mostly, thermal power plants produce bulk quantities of ash waste [37]. Efforts are made to get new novelties either through recycling or integrating waste into innovative products. For this work, bulk quantity of raw materials and high temperatures are essential to achieve novel attempts. One such way is to use wastes for making products like ceramic. Coal ashes are produced as by-products during combustion of coal in thermal power-plants [38, 39]. Both source and burning process affect physical and chemical properties of coal ashes [40]. Elements present in coal ashes are silicon, calcium, aluminum, iron, magnesium, sulfur, carbon, which is in major percentage [41]. In three ways, coal ashes are collected. They are pond ash, bottom ash, and fly ash [42]. Pond ash is one of the wastes obtained from burning of coal in boilers. It is mainly obtained from wet disposal of fly ash, which when gets mixed with bottom ash during dispose in large pond or dykes (as slurry). Pond ash is being generated in an alarming rate. Generation of pond ash is posing a threat to our environment and the management is now experiencing problem to dispose it off. This has become a thrust area for engineering research [43].

To enhance electrical and physico-mechanical properties of pond ash, it is essential to prepare composite. Some reports are available on industrial waste based composite. Usually, industrial wastes contain different type of elements (i.e., Cu, Pb, Cd, Ag, Mo, Fi, Ti, Na, Mn, S, P, Zn, and Cl) in different proportions. Some of these elements have insulating, while others are semiconducting in nature [44, 45]. Hence, one way is to explore possibilities of using waste for making electronic materials (insulators and semiconductors).

Dielectric properties and DC resistivity of industrial waste based materials have been studied in the range of temperature (300-500 °C) [46].

Thermal power plant ash mainly comprises of crystalline phases and glassy phase [47]. Work is going on to improve dielectric and electrical insulation properties of industrial wastes [48–51].

DC resistivity, dielectric constant, and dielectric loss are main considerations for fabricating insulators [52].

The current chapter reports results of investigation on composite materials preparation, phase, topography, chemical groups, and electrical resistance analyses at room temperature. Purpose is to depict and interpret properties of composite materials.

8.2 Experimental Section

8.2.1 Materials and Chemicals

The main ingredient i.e., pond ash is collected from National Aluminium Company (NALCO) Odisha, India. Pyrophyllite is used as an additive and is procured from Merck India. Dextrin $(C_6H_{10}O_5)_n \cdot xH_2O)$ is required to bind materials. It is also obtained from Merck India. 6% water is added to the above mixture to prepare green paste. Compositions of pond ash and pyrophyllite are shown in Table 8.1. Optical photographs of three ingredients i.e., pond ash, dextrin, and pyrophyllite are shown in Figure 8.1.

8.3 Preparation of PA/PY Composite Materials

PA/Pyrophyllite composite is prepared by solid-state sintering process. Compositions (%) of Pond ash and PY materials are shown in Table 8.2. Three steps involved are described below [55, 56];

Step-1 Processing of raw materials (i.e., pond ash)
Collected pond ash is grinded in a ball mill (for 5 h) to reduce particle sizes. Such powder is screened through 240 mesh. Sieved particles are dried in a heating oven at 120 °C (2 h). Purpose of heating sieved pond ash is to eliminate moisture present in it. Such ash is designated as processed pond ash.

Step-2 Making of Green Pellets
Appropriate quantity of processed pond ash (i.e., 50%) is kept in a motar with pastel. Requisite amounts of pyrophyllite (i.e., 50%) is added to pond ash contained in mortar and pastel. They are grinded to very fine particles. Small quantity of dextrin (i.e., 0.05 %) is mixed with above mixture and then, grinded further for half an hour in order to mix them homogeneously. Small amount of water is added to make the mixture pasty. This paste samples are termed as green paste. Paste is ready to prepare pellet.

For making green pellet, universal testing machine (UTM) with pelletizer is employed. Different parts of pelletizer are cleaned, washed, and dried before use. Required quantity of paste materials is kept in cleaned pelletizer. It is compacted in an universal testing machine by applying a

Table 8.1 Constituents with percentage (%) of processed pond ash and pyrophyllite.

S. ID	Compositions (%)												Ref.
	Fe_2O_3	Al_2O_3	SiO_2	P_2O_5	Cr_2O_3	ZnO	MgO	C	CaO	TiO_2	MnO	LOI	
PA	3.85	28.3	62.8	0.32	0.04	0.027	0.049	1.15	0.7	1.84	0.03	0.5	[53]
PY	---	37.66	44.80	---	---	---	T	---	0.50	0.60	---	14.33	[54]

Note: S. ID=Sample ID, PA=Pond Ash, PY=Pyrophyllite, T=Trace.

Figure 8.1 Optical image of pond ash (a), Dextrin (b) and pyrophyllite (c).

Table 8.2 Compositions (%) of pond ash/PY materials preparation.

	Individual composition name	
S. no.	Pond ash	Pyrophyllite (PY)
1	50	50
2	60	40
3	70	30
4	80	20
5	90	10

load of 10 MPa. Load is maintained for five minutes to complete compaction. Then, pelletizer is demoulded and is collected pellets from pelletizer. Dimensions (Thicknesses and diameter) of pellet(s) are measured. Such pellet is ready for sintering.

Step-3 Sintering process
Green pellet samples are kept in a muffle furnace. They are bisquetted at 900 °C for 1 h. This process has helped to eliminate intacted water and binder. Bisquetted pellet(s) are sintered at 1200 °C for 2 h. Dimensions (i.e., thickness and diameter) of sintered pellet are found to be 0.5 cm and 3 cm, respectively. Entire process is shown schematically in Figure 8.2. Sintered pellets are ready for different characterizations process.

Similar process is followed in other batches.

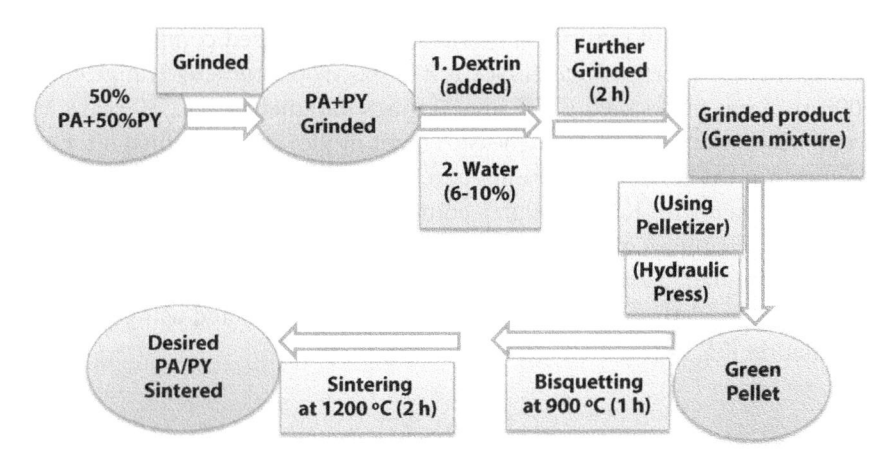

Figure 8.2 Flow chart for preparation of sintered PA/PY (50:50) composite materials.

8.4 Test Methods

X-ray diffraction analyses (XRD) are made to identify different phases. For this purpose, Phillips PW-1710 advanced wide angle X-ray diffractometer attached with Phillips PW-1729 X-ray generator is used. CuKα radiation ($\lambda \sim 0.154$ nm) is used for analysis. Pellet sample is put on a sample holder (made of quartz) to obtained XRD pattern in X-ray machine. Measurement is performed at room temperature. Sample is scanned between 20° to 70° with scanning rate of 2°/min.

Optical image is taken during hardness measurement by nano indentation (Anton Paar-NHT, Switzerland). Pellet sample of pond ash/pyrophyllite (60:40) composite is fixed on the moving sample holder.

Morphological analyses are done by Carl Zeiss Supra 40 field emission scanning electron microscope (FESEM). Sintered samples are Gold coated before being placed in FESEM. Sputtering technique is used for coating the sintered samples with gold. During test, operating voltage is kept at 30 kV. EDS analyses are performed together with FESEM microscopic analyses.

TEM (5022/22, TecnaiG220 S-Twin, Czech Republic) technique is used to investigate image and SAD pattern of raw pond ash (PA) powder and pond ash/pyrophyllite (60:40) composite. TEM specimens of pond ash and pond ash/pyrophyllite (60:40) composite are prepared by microtone technique (LEICA Microsystem, GmBH, A-1170). Then, specimens are transferred to Cu TEM grids.

FTIR spectrum of pyrophyllite, and sintered PA based composite samples are noted. Thermo Nicolet Nexus 870 spectrophotometer (range 400-4000cm^{-1}) has performed FTIR test. Settings parameters (i.e., 50 scans at 4 cm^{-1} resolution, Absorbance measurement mode) are kept constant. For performing the test, pellets are prepared using small amount of sintered samples mixed with KBr through a compression molding system. A background spectrum is collected before test. Then, pellet is put in a FTIR sample holder and data are collected.

Electrical resistivity of sintered pellet sample is estimated using following relation [55, 56];

$$\rho = \frac{R \times A}{l} \quad (8.1)$$

Here, A is area of cross – section of electrode $= \frac{\pi}{4} D^2 (\text{centimeter}^2)$, R is resistance of sintered samples (MΩ), l is thickness of sample in centimeter (cm), ρ is the resistivity of samples in Ω.cm. For this measurement, Multimeter is used.

8.5 Results and Discussion

Resistivity values of pond ash and pond ash/pyrophyllite composite in different proportions (sintered at 1200 °C) are shown in Table 8.3. It is evident that Pond ash/pyrophyllite composite in 60:40 has exhibited the highest resistivity i.e 38.587185 × 10^9 (Ω.cm) as against resistivity value of sintered pond ash (i.e., 8.19801 × 10^9 Ωcm). This is explained to be due to formation of mullite phase in different quantities [55, 56].

FTIR absorption bands of pond ash are well studied [57]. Presence of different absorption stretching frequency of pond ash powder corresponds to pond ash components. Stretching frequency at 600 cm^{-1} matches with Si-O-Al band [58]. Stretching frequencies at 1098 and 1608 cm^{-1} are accredited to Si-O-Si asymmetric band [57] and H-O-H [59], respectively.

FTIR spectrum of pond ash (60%)/pyrophyllite (40%) composite sintered at 1200 °C is shown in Figure 8.3. Figure 8.3 shows occurrence of various types of absorption bands arising at different wave numbers i.e., 3441, 2918, 1823, 1638, 1117 and 907 cm^{-1}. Stretching bands at 3441 cm^{-1} and 1638 cm^{-1} indicate vibrations of O-H stretching bands [60] and H-O-H bending bands [59], respectively. Absorption peak at 1117 cm^{-1} is

Table 8.3 Electrical resistivity value of pond ash/pyrophyllite (PA/PY) based sintered samples [radius=1.5 cm and thickness=0.5 cm].

S. no.	Sample ID	Sintering temperature (°C)	Electrical resistance (Giga. Ohm, GΩ)	Resistivity (Ω.cm)
1	S1	1200	0.58	8.19801×10^9
2	S2	1200	1.43	20.212335×10^9
3	S3	1200	2.73	38.587185×10^9
4	S4	1200	1.45	20.495825×10^9
5	S5	1200	1.13	15.971985×10^9
6	S6	1200	0.91	12.862395×10^9

*Note: S1-pond ash sintered, S2-pond ash/pyrophyllite (50/50) sintered at 1200 °C, S3-pond ash/pyrophyllite (60/40) sintered at 1200 °C, S4-pond ash/pyrophyllite (70/30) sintered at 1200 °C, S5-pond ash/pyrophyllite (80/20) sintered at 1200 °C, and pond ash/pyrophyllite (90/10) sintered at 1200 °C.

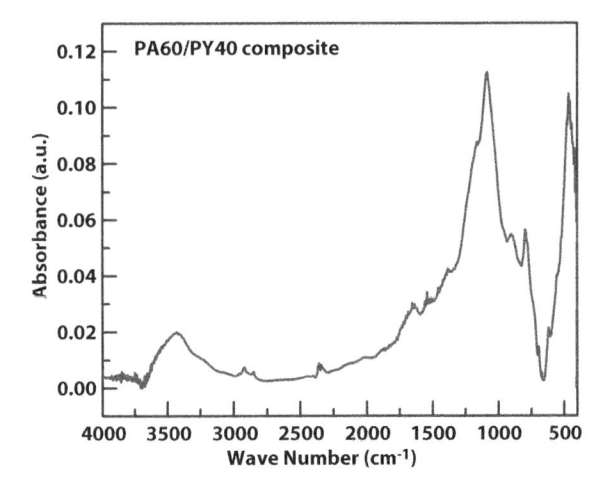

Figure 8.3 FTIR spectrum of pond ash (60%)/pyrophillite (40%) composite.

shown by Si-O band [61] and it signifies presence of silicate groups. $Al^{3+}O^{2-}$ absorption band is also seen at 907 cm^{-1} [62].

Figure 8.4a shows the XRD pattern of pond ash. Identified different phases are alumina, silica, and mullite. The highest intense peak is due to silica phase. Other two predominant phases are mullite and alumina (Figure 8.4a) [55, 56].

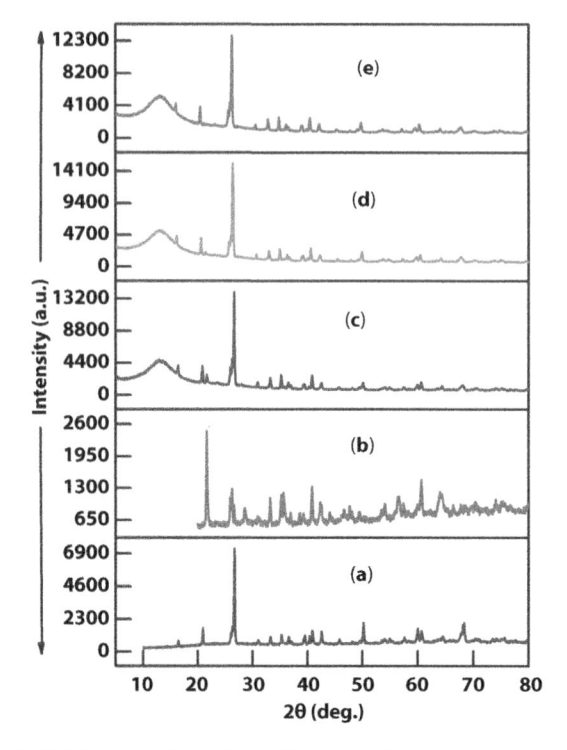

Figure 8.4 Pond ash (a), sintered pond ash (b), pond ash-pyrophillite (60-40 (c)), pond ash-pyrophillite (d) (70-30), pond ash-pyrophillite (e) (80-20).

Figures 8.4b, 8.4c, 8.4d, and 8.4e are XRD pattern of pond ash and pond ash/pyrophyllite composite with different proportions, which are sintered at 1200 °C for 1 h. Details of predominant phases are shown in Table 8.4. Table 8.4 specifies crystal system, space group, space group number, cell dimension, cell angle, density of unit cell, etc. It can be observed that all sintered samples show presence of mullite phase. In the pattern, Different amount of mullite phase can be observed. Amount of mullite phase depends on silica and alumina conversion, which are present in both pond ash and pyrophyllite components. Alumina and silica binary phase diagrams suggest that mullite phase dominates at 1200 °C [63].

Optical image of pond ash (60%)/pyrophillite (40%) composite sintered at 1200 °C is shown in Figure 8.5. Presences of white patches in different shapes are dispersed in the matrix. Distinctive feature is that they largely vary in their aspect ratios. Optical micrograph shows poor contrast and therefore FESEM study is carried out to understand these features in more details.

Table 8.4 Structural information of pond ash/pyrophillite composite with different proportions.

Crystal ID	Mineral name	Crystal system	Space group	Space group number	Cell dimension	Crystal angle	Calculated density (g/cm^3)
As-received PA	Quartz SiO_2	Hexagonal	P3221	154	a=b≠c	α=β=90°, γ=120°	2.65
PA 1200 °C	Mullite $Al_2(Al_{2.5}Si_{1.5})O_{9.75}$	Orthorhombic	Pbam	55	a≠b≠c	α=β=γ=90°	3.17
PA 1200 °C	Quartz SiO_2	Cubic	Fd-3m	227	a=b=c	α=β=γ=90°	2.21
PA+PY (60%+40%) 1200 °C	Mullite $Al(Al_{1.272}Si_{0.728}O_{4.864})$	Orthorhombic	Pbam	55	a≠b≠c	α=β=γ=90°	3.16
PA+PY (60%+40%)	Quartz SiO_2	Hexagonal	P3221	154	a=b≠c	α=β=90°, γ=120°	---
PA+PY (70%+30%) 1200 °C	Mullite $Al_{4.75}Si_{1.25}O_{9.63}$	Orthorhombic	Pbam	55	a≠b≠c	α=β=γ=90°	3.13

(Continued)

Table 8.4 Structural information of pond ash/pyrophillite composite with different proportions. (*Continued*)

Crystal ID	Mineral name	Crystal system	Space group	Space group number	Cell dimension	Crystal angle	Calculated density (g/cm^3)
PA+PY (70%+30%) 1200 °C	Quartz SiO$_2$	Hexagonal	P3221	154	a=b≠c	α=β=90°, γ=120°	---
PA+PY (80%+20%) 1200 °C	Silicon dioxide	Hexagonal	P3121	152	a=b≠c	α=β=90°, γ=120°	2.63
PA+PY (90%+10%) 1200 °C	Potassium Hydrogen Phosphide KPH$_2$	Monoclinic	P21/n	14	a≠b≠c	α=γ≠β	1.44

*Note: PA-Pond ash, PY-Pyrophyllite, SiO$_2$-Silicon dioxide, KPH$_2$-Potassium hydrogen phosphide.

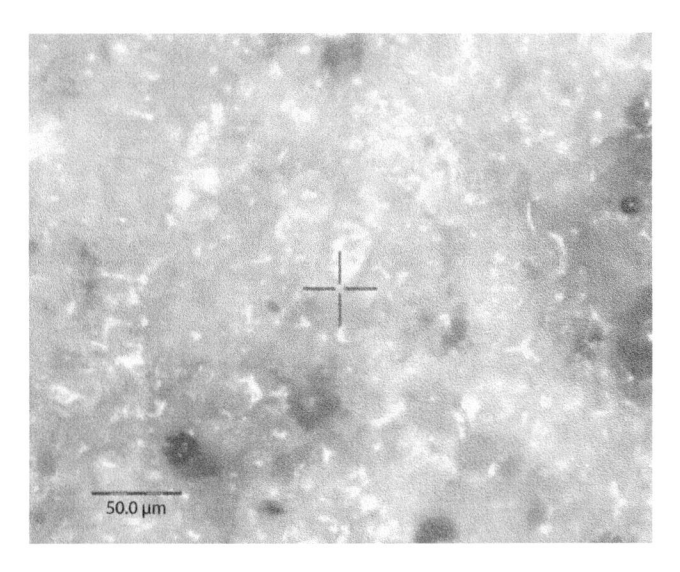

Figure 8.5 Pond ash (60%)/pyrophillite (40%) composite.

FESEM images of pond ash and pyrophyllite are shown in Figures 8.6a and 8.6b. FESEM image of PA constitutes ellipsoid particles, widely varying in sizes. Aspect ratio of pond ash particles is 2:1. Major axis of the largest particles is measured to be around 12 micron and the minor axis is around 6 micron. Figure 8.6b shows micrograph of pyrophyllite. The microstructure is flaky in nature having different sizes. This structure shows preferred orientation direction.

On sintering of composite of pond ash (60%)/pyrophyllite (40%), there is appreciable change in microstructure. At lower magnification (Figure 8.6c), microstructure of PA/PY composite indicates non-uniform segregated structure with small pores. This is due to interaction of pyrophyllite (PY) with PA particles. The microstructure contains crystalline phases. Higher magnification (Figure 8.6d) indicates fused mass which is formed due to the interaction of pyrophyllite (PY) phase and PA phase. Crystalline phase formation is presumably of mullite phase as is detected by XRD analyses (Figure 8.4).

EDS analyses are done for three sintered samples during FESEM studies (Figure 8.7). O, Al, Si, Ti, Fe, K, Na, Ca, and Mg elements occur in the pond ash/pyrophyllite sintered composite (Figure 8.7). Ti, K, and Na elements are present in trace amounts in the composite (Figure 8.7). All the elements are shown in Figure 8.7 are well an agreement with XRD results.

Figure 8.8 shows Transmission electron micrographs of raw pond ash and Pond ash/pyrophyllite (60:40) composite. They are taken at high

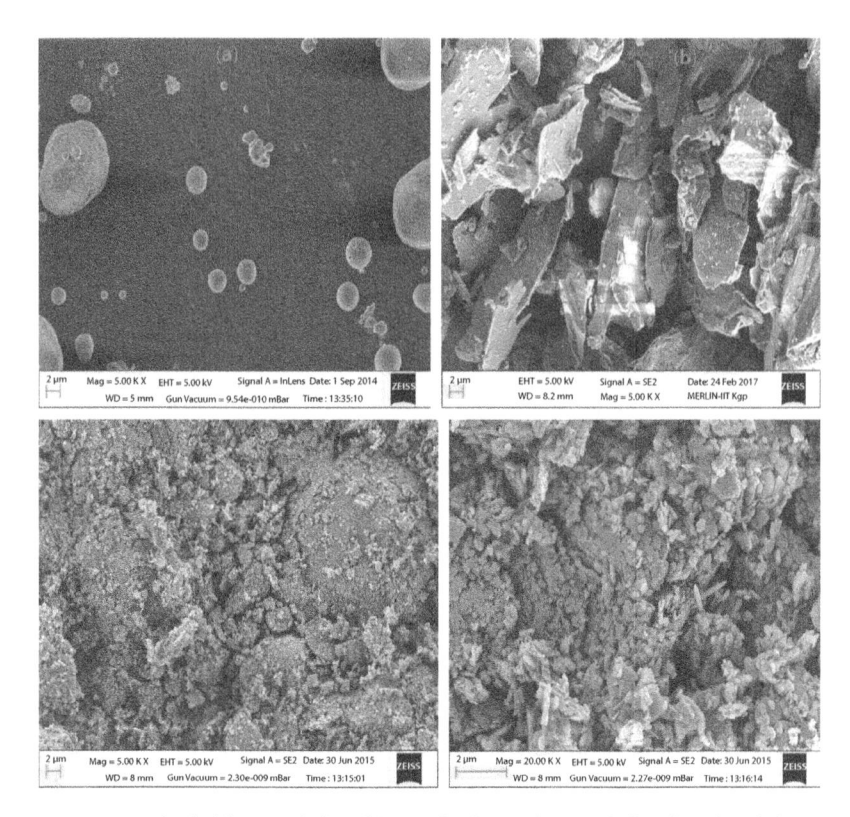

Figure 8.6 Pond ash (a), pyrophillite (b), pond ash (60%)/pyrophillite (40%) with lower magnification (c), and pond ash (60%)/pyrophillite (40%) with lower magnification (d).

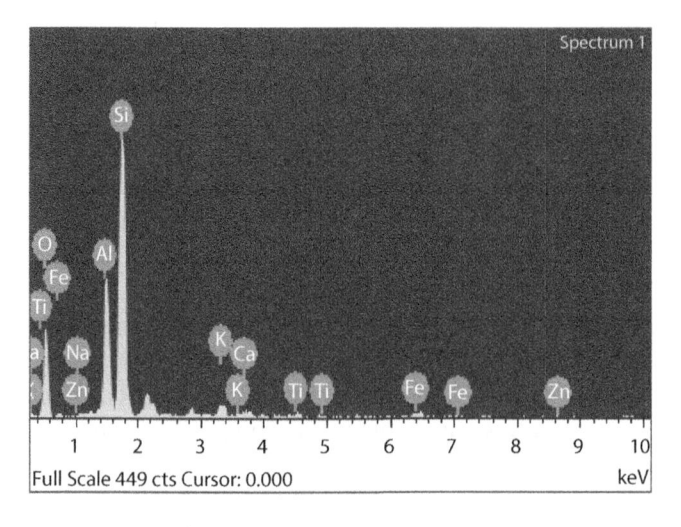

Figure 8.7 EDS spectrum of pond ash/pyrophyllite (60:40) composite sintered at 1200 °C.

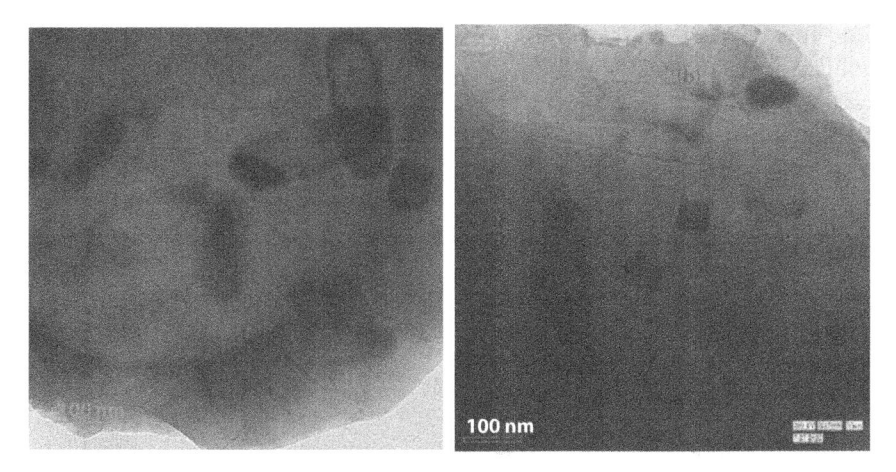

Figure 8.8 TEM image of raw pond ash (a) and PA/PY (60:40) composite sintered at 1200 °C (b).

magnification. The contrast of the picture (Figure 8.8a) suggests three dimensional shapes of embeded crystalline phases. Similarly, TEM micrograph of pond ash/pyrophyllite (60:40) composite in Figure 8.8b shows a fewer shadows of a few crystalline phases appeared in the matrix. It can be inferred from the Figure 8.8b that amount of crystalline phase is reduced by sintering (1200 °C) of pond ash and pyrophyllite.

Electron diffraction pattern i.e., SAD pattern of both the materials are shown in Figures 8.9a and 8.9b. For pond ash (Figure 8.9a), there are many bright spots appearing in different circumference of circles. These spots

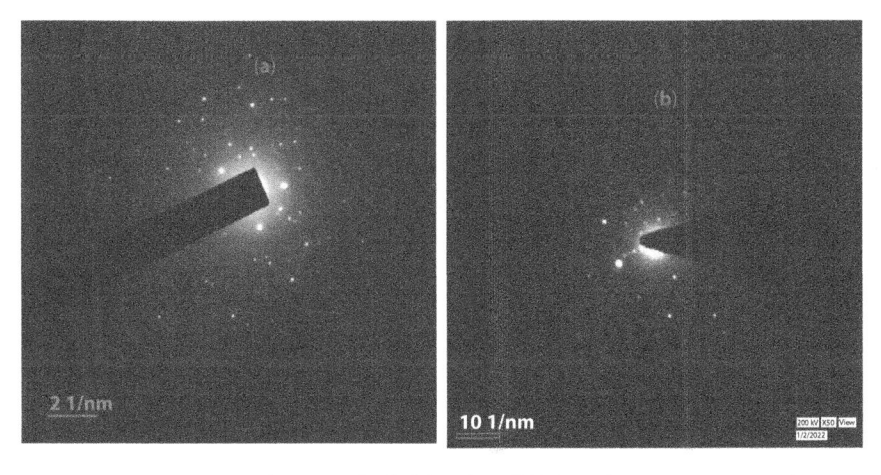

Figure 8.9 SAD pattern of raw pond ash (a) and PA/PY (60:40) composite sintered at 1200 °C (b).

are superimposed pattern from polycrystalline phases present in sintered pond ash. However, sintered PA/PY (60:40) composite (Figure 8.9b) shows a fewer bright spots appearing in the diffraction pattern. These spots are presumably due to electron diffraction of crystalline phases present in the sintered products. FESEM and XRD analyses suggest that crystalline phases are reduced in sintered composite due to formation of glassy phase at high sintering temperature. Predominant phase is identified i.e., mullite phase for sintered pond ash and pond ash/pyrophyllite composite via XRD analyses.

8.6 Conclusions

The main aim of this work is to develop a composite material having high electrical resistivity. For this, pond ash is chosen as a primary raw material, which is a waste from thermal power plant. Addition of pyrophyllite to pond ash increased alumina and silica ratios in the mixture. Judicial adjustment of composition and with optimum treatment combination, it has been possible to obtain a product which has acquired a resistivity value of 38.587185 ohmcm. FESEM image of pond ash shows ellipsoids particles having 2:1 aspect ratios. They are widely varying in sizes. Pyrophyllite shows flakey-structure. Due to sintering of pond ash and pyrophyllite in 60:40 ratios, there is marked change in microstructure. Occurrence of both crystalline and glassy phases is observed. TEM micrograph of sintered pond ash and PA/PY (60:40) composite show marked difference. Prominent TEM crystalline phases are observed in SAD pattern of sintered pond ash. However, quantity in crystalline phases in sintered PA/PY (60:40) composite due to interaction between two raw materials has decreased. Elements present in the samples are identified by EDS analyses. Mullite and quartz phases are clearly identified from XRD pattern of PA/PY composite sintered at 1200 °C. Crystal structures of mullite and quartzite phase are orthorhombic and tetragonal (JCPDS file). Densities of the orthorhombic and tetragonal unit cell are 3.16 g/cm^3 and 2.26 g/cm^3, respectively. In FTIR spectra, Si-O-Al- band and Si-O band are indicated. These bands designate the formation of mullite and quartz phases. It is in good agreement with XRD analyses.

Acknowledgements

First author would like to thank Prof. Munesh Chandra Adhikary, PG Council Chairman, Fokir Mohan University, for his invaluable guidance, advice and constant inspiration throughout the entire program. First author also thanks Mr. Mukteswar Mohapatra, Fokir Mohan University for his support. The authors convey their sincere thanks to GIET, University Gunupur, Rayagada, Odisha, India for providing Lab facilities to do the research work. Authors would also like to thank the CRF, IIT Kharagpur for providing testing facilities.

References

1. J. C. Egbai, (2013), Kaolin Quantification in Ukwu-nzu and Ubulu-uku using Electrical Resistivity Method, *Conference Proceedings International Journal of Recent Research and Applied Studies*, Vol. 14.
2. O. A. Babalola and T. Akomolafe, (2008), Effects of Kaolin Particle Size and Annealing Temperature on the Resistivity of Zinc-Kaolin Composite Resistors, *Journal Applied Science Environmental Management*, Vol. 12, 113.
3. L. Li and Y. Li, (2017), Development and trend of ceramic cutting tools from the perspective of mechanical processing, *IOP Conf. Series: Earth and Environmental Science*, Vol. 94, pp. 1-6.
4. M. Zhu, R. Ji, Z. Li, H. Wang, L. L. Liu, and Z. Zhang, (2016), Preparation of Glass Ceramic Foams for Thermal Insulation Applications from Coal Fly Ash and Waste Glass, *Construct. and Building Materials*, Vol. 112, pp. 398-405.
5. H. E. Exner and E. Arzt, (1996), Sintering Processes, In *Physical Metallurgy*, Ed. by R. W. Cahn and P. Haasen, 4th Ed. Elsevier Science, Amsterdam, Vol. 3, pp. 2628-2662.
6. W. V. Siemens, (1966), Inventor and Entrepreneur: Recollections of Werner Von Siemens, London, England.
7. T. Tanaka and T. Imai, (2013), Advances in Nanodielectric Materials Over the Past 50 years, *IEEE Electrical Insulation Magazine*, Vol. 1, pp. 10-23.
8. T. Tanaka, (2005), Dielectric nanocomposites with insulating properties, *IEEE Transactions on Dielectrics and Electrical Insulation*, Vol. 12, pp. 914-928.
9. E. C. Nzenwa, A. D. Adebayo, (2019), Analysis of Insulators for Distribution and Transmission Networks, *American Journal of Engineering Research (AJER)*, Vol. 8, pp. 138-145.
10. C. L. Goldsmith, A. Malczcwski, J. J. Yao, S. Chen, J. Ehmk, D. H. Hinzel, (1999), RFMEMS-Based Tunable Filters. *Int. J.RF Microwave CAE*, Vol. 9, 362.
11. G. Subramanyam, F. V. Keuls, and F. A. Miranda, (1998), Novel K-band Tunable Microstrip v-band Pass Filter using Thin Film HTS/Ferroelectric/Dielectric Multilayer Configuration. *IEEE Microwave Guided Wave Letter*, Vol. 8, 78.
12. E. Martini, A. Pavese, D. Tabacchi, D. M. Fortuna, and A. Fortuna, (2021), Effects of sintering temperature on microstructure and properties of

sanitaryware ceramic produced with waste material, *Cerâmica*, Vol. 67, pp. 39-47, http://dx.doi.org/10.1590/0366-69132021673813022.
13. Laura Silvestroni and Diletta Sciti, Sintering Behavior, (2010), Microstructure, and Mechanical Properties: A Comparison among Pressureless Sintered Ultra-Refractory Carbides, *Advances in Materials Science and Engineering* Vol. 2010, pp. 1-11.
14. H. Olphen, (1977), An Introduction to Clay Colloid Chemistry, John Wiley & Sons, New York.
15. T. K. Mukhopadhyay, S. Ghatak, and H. S. Maiti, (2010), Pyrophyllite as Raw Material for Ceramic Applications in the Perspective of its Pyro-Chemical Properties, *Ceram. Int.*, Vol. 36, pp. 909-916.
16. D. Perkins, (1998), Mineralogy, Prentice-Hall Inc, Upper Saddle River, New Jersey 07458.
17. K. J. MacKenzie, S. Komphanchai, and R. Vagana, (2008), Formation of Inorganic Polymers (Geopolymers) from 2:1 Layer Lattice Aluminosilicates, *J. Eur. Ceram. Soc.*, Vol. 28, pp. 177-181.
18. S. Mohammadnejad, J. L. Provis, and J. S. van Deventer, (2014), Effects of Grinding on the Preg-Robbing Behaviour of Pyrophyllite, *Hydrometallurgy*, Vol. 146, pp. 154-163.
19. J. Oshitani, M. Kondo, H. Nishi, and Z. Tanaka, (2003), Separation of Silicastone and Pyrophyllite by a Gas-Solid Fluidized Bed Utilizing Slight Difference of Density, *Adv. Powder Technol.*, Vol. 14, pp. 247-258.
20. J. Zhang, L. Hu, R. Pant, Y. Yu, Z. Wei, and G. Zhang, (2013), Effects of Interlayer Interactions on the Nanoindentation Behavior and Hardness of 2:1 Phyllosilicates, *Appl. Clay Sci.*, Vol. 80, pp. 267-280.
21. D. Zhang, C. Zhou, C. Lin, D. Tong, and W. Yu, (2010), Synthesis of Clay Minerals, *Appl. Clay Sci.*, Vol. 50, pp. 1-11.
22. D. Kibanova, M. Trejo, H. Destaillats, and J. Cervini-Silva, (2011), Photocatalytic Activity of Kaolinite, *Catal. Commun.*, Vol. 12, pp. 698-702.
23. L. V. Barbosa, L. Marçal, E. J. Nassar, P. S. Calefi, M. A. Vicente, R. Trujillano, V. Rives, A. Gil, S. A. Korili, and K. J. Ciuffi, (2014), Kaolinite-Titanium Oxide Nanocomposites Prepared Via Sol-Gel as Heterogeneous Photocatalysts for Dyes Degradation, *Catal. Today*, Vol. 246, pp. 133-142.
24. L. Xia, H. Zhong, G. Liu, Z. Huang, Q. Chang, and X. Li, (2009), Comparative Studies on Flotation of Illite, Pyrophyllite and Kaolinite with Gemini and Conventional Cationic Surfactants, *Trans. Nonferr. Met. Soc. China*, Vol. 19, pp. 446-453.
25. H. Zhong, G. Liu, L. Xia, Y. Lu, Y. Hu, S. Zhao, and X. Yu, (2008), Flotation Separation of Diaspore from Kaolinite, Pyrophyllite and Illite Using Three Cationic Collectors, *Miner. Eng.*, Vol. 21, pp. 1055-1061.
26. L. Xia, H. Zhong, and G. Liu, (2010), Flotation Techniques for Separation of Diaspore from Bauxite using Gemini Collector and Starch Depressant, *Trans. Nonferr. Met. Soc. China*, Vol. 20, pp. 495-501.

27. Y. Hu, X. Liu, and Z. Xu, (2003), Role of Crystal Structure in Flotation Separation of Diaspore from Kaolinite, Pyrophyllite and Illite, *Miner. Eng.* Vol. 16, pp. 219-227.
28. M. Erdemoğlu and M. Sarıkaya, (2002), the Effect of Grinding on Pyrophyllite Flotation, *Miner. Eng.* 15 pp. 723-725.
29. Y. Hu and X. Liu, (2003), Chemical Composition and Surface Property of Kaolins, *Miner. Eng.* Vol. 16, pp. 1279-1284.
30. S. K. Jena, S. Singh, D. S. Rao, N. Dhawan, P. K. Misra, and B. Das, (2015), Characterization and Removal of Iron from Pyrophyllite Ore for Industrial Applications, *Miner., Metall. Process*, Vol. 32 pp. 102-110.
31. F. K. Crundwell, (2016), on the Mechanism of The Flotation of Oxides and Silicates, *Miner. Eng.* Vol. 95, pp. 185-196.
32. W. Stumm, (1992), Chemistry of the Solid-Water Interface: Processes at the Mineral water And Particle-Water Interface in Natural Systems, *Soil Sci.* Vol. 156, pp. 21-72.
33. D. Wang and Y. Hu, (1988), Solution Chemistry of Flotation in, Hunan Science and Technology Press, Beijing, China.
34. R.J. Hunter, (2013), Zeta Potential in Colloid Science: Principles and Applications, Academic Press.
35. S. B. Johnson, G. V. Franks, and P. J. Scales, (2000), Surface Chemistry-Rheology Relationships in Concentrated Mineral Suspensions, *Int. J. Miner. Process.* Vol. 58, pp. 267-304.
36. K. G. Bhattacharyya and S. S. Gupta, (2008), Adsorption of a Few Heavy Metals on Natural and Modified Kaolinite and Montmorillonite: A Review, *Adv. Colloid Interface Sci.* Vol. 140, pp. 114-131.
37. Manas Ranjan Senapati, (2011), Fly ash from thermal power plants – waste management and overview, Vol. 100, pp. 1791-1794, https://www.jstor.org/stable/2407754.
38. K. S. Sajwan, T. Punshon, and J. C. Seaman (2006), Production of Coal Combustion Products and Their Potential Uses, In: Sajwan, K.S., Twardowska, I., Punshon, T., Alva, A.K. (eds) *Coal Combustion Byproducts and Environmental Issues*. Springer, New York, NY.
39. R. S. Kalyoncu and D.W. Olson, (2001), Coal Combustion Products: U.S. Geological Survey, Coal Combustion Products, pp. 1-12.
40. M. Farhad Howladar and Md. Raisul Islam, (2016) A study on Physico-Chemical properties and Uses of Coal Ash of Barapukuria Coal Fired Thermal Power Plant, Dinajpur, for Environmental sustainability, *Energy, Ecology and Environment*, Vol. 1, pp. 233–247, https://doi.org/10.1007/s40974-016-0022-y.
41. M.L.D. Jayaranjan, E.D. van Hullebusch, and A.P. Annachhatre, (2014), Reuse options for coal fired power plant bottom ash and fly ash, Rev. Environ. Sci. Biotechnol. Vol. 13, pp. 467–486, https://doi.org/10.1007/s11157-014-9336-4.
42. https://cen.acs.org/articles/94/i7/New-Life-Coal-Ash.html.

43. https://www.thethirdpole.net/en/climate/coal-ash-is-a-serious-hazard-to-our-health-and-the-environment/
44. O. Popov, A. Iatsyshyn, V. Kovach, V. Artemchuk, I. Kameneva, O. Radchenko, K. Nikolaiev, V. Stanytsina, A. Iatsyshyn, and Y. Romanenko, (2021), Effect of Power Plant Ash and Slag Disposal on the Environment and Population Health in Ukraine, *J Health Pollut.*, Vol. 11, pp. 1-10.
45. M. S. Chavali and M. P. Nikolova, (2019), Metal oxide nanoparticles and their applications in nanotechnology, *SN Applied Sciences*, Vol. 1, pp.1-30, https://doi.org/10.1007/s42452-019-0592-3.
46. E. P. A. Chinese, (1996), China Identification Standard for Hazardous Waste Toxicity (GB5085.3-1996), Chinese Environmental Science Press, Beijing.
47. S. B. Kondawar, A. D. Dahegaonkar, V. A. Tabhane, and D. V. Nandanwar, (2014), Thermal and Frequency Dependence Dielectric Properties of Conducting Polymer/Fly Ash Composites, *Advanced Materials Letters*, Vol. 5, 360.
48. A. Pattnaik, S. K. Bhuyan, S. K. Samal, A. Behera, and S. C. Mishra, (2016), Dielectric Properties of Epoxy Resin Fly Ash Composite, *IOP Conference Series: Materials Science and Engineering*, Vol. 115, 1.
49. M. Mishra, A. P. Singh, and S. K. Dhawan, (2013), Utilization of Fly Ash-A Waste By-product of Coal for Shielding Application, *Journal of Environmental Nanotechnology*, Vol. 2, 74.
50. J. P. Dhal and S. C. Mishra, (2013), Investigation of Dielectric Properties of a Novel Hybrid Polymer Composite using Industrial and Bio-waste, *Journal of Polymer Composites*, Vol. 1, 22.
51. K. Singh, T. Quazia, S. Upadhyaya, and P. Sakharkarb, (2005), Development of Low Permittivity Material using Fly Ash, *Indian Journal of Engineering & Materials Sciences*, Vol. 12, 345.
52. D. E. Yıldız and I. Dokme, (2011), Frequency and Gate Voltage Effects on the Dielectric Properties and Electrical Conductivity of $Al/SiO_2/p$-Si Metal-Insulator-Semiconductor Schottky Diodes, *Journal of Applied Physics*, Vol. 110, 014507.
53. M. K. Panigrahi, R. R. Dash, R. I. Ganguly, (2018), Development of Novel Constructional Material From Industrial Solid Waste as Geopolymer for Future Engineers, *IOP Conference Series: Materials Science and Engineering*, Vol.410, pp.1-12.
54. X. Liu, and M. Bai, (2017) Effect of chemical composition on the surface charge property and flotation behavior of pyrophyllite particles, *Advanced Powder Technology*, Vol. 28, pp. 836-841, http://dx.doi.org/10.1016/j.apt.2016.12.009.
55. M.K. Panigrahi, (2022), Investigation of Structures of Sintered Fly Ash Materials: Resources of Industrial Wastes, *Bulletin of Scientific Research*, Vol. 4, pp.1-10.

56. M.K. Panigrahi, (2021), Investigation of Structural, Morphological, Resistivity of Novel Electrical Insulator: Industrial Wastes, *Bulletin of Scientific Research*, Vol. 3, pp. 51-58.
57. A. Jose, M. R. Nivitha, M. Krishnan, and R. G. Robinson, (2020), Characterization of cement stabilized pond ash using FTIR spectroscopy, *Construction and Building Materials*, Vol. 263, 120136.
58. Y. Liu, F. Zeng, B. Sun, P. Jia, and I. T. Graham, (2019), Structural Characterizations of Aluminosilicates in Two Types of Fly Ash Samples from Shanxi Province, North China, *Minerals*, Vol. 9, pp. 358-16.
59. P. Rożek, M. Król, and W. Mozgawa, (2018), Spectroscopic studies of fy ash-based geopolymers, *Spectrochim Acta - Part A Mol Biomol Spectrosc* Vol. 198, pp. 283-289, https://doi.org/10.1016/j.saa.2018.03.034.
60. S. Kumar, F. Kristály, and G. Mucsi, (2015), Geopolymerisation behaviour of size fractioned fly ash, *Adv Powder Technol*, Vol. 26, pp. 24-30, https://doi.org/10.1016/j.apt.2014.09.001.
61. Z. Kledyński, A. Machowska, B. Pacewska & I. Wilińska, (2017), Investigation of hydration products of fly ash-slag pastes, *Journal of Thermal Analysis and Calorimetry*, Vol. 130, pp. 351-363, https://doi.org/10.1007/s10973-017-6233-4.
62. D. Bondar and R. Vinai, (2022), Chemical and Microstructural Properties of Fly Ash and Fly Ash/Slag Activated by Waste Glass-Derived Sodium Silicate, *Crystals*, Vol. 12, pp. 913-12, https:// doi.org/10.3390/cryst12070913.
63. R. Roy, D. Das, and P. K. Rout, (2022), A Review of Advanced Mullite Ceramics, *Engineered Science*, Vol. 18, pp. 20-30.

9

High Resistance Sintered Pond Ash/k-Feldspar (PA/k-FD) Ceramics

M. K. Panigrahi[1*], R.I. Ganguly[2] and R.R. Dash[3]

[1]Department of Materials Science, Maharaja Sriram Chandra Bhanja Deo University, Balasore, Odisha, India
[2]Department of Metallurgical Engineering, National Institute of Technology, Raurkela, Odisha, India
[3]CSIR-National Metallurgical Laboratory, Jamshedpur, Jharkhand, India

Abstract

Pond ash is obtained from ponds near a thermal power station where burnt ash of coal is dumped. Due to storing of ash under water for a long-time, there are meta-morphological changes in ash. Chemical and mineralogical compositions of pond ash constitute alumino-silicate compounds. This is present in a major quantity with a few others such as TiO_2, Fe_2O_3, etc. Present study has attempted to utilize pond ash and it's composite for making cheaper high resistance insulators. To achieve this, Pond ash has been mixed with k-feldspar. This is sintered at high temperature. Product thus obtained is characterized by X-ray diffraction (XRD) method, Fourier transformation infra-red spectroscopy (FTIR), optical microscopic studies, field emission scanning electron microscopic studies (FESEM), and transmission electron microscopic (TEM) studies. Resistivity measurement is done. Raw materials are also characterized for identification of phases, chemical group, using XRD, FTIR, and optical microscope/FESEM/TEM equipments. Quartz, mullite, and other phases are identified by XRD analyses. For identification of elements, EDS analyses are made for sintered pond ash/k-feldspar composite which has the highest electrical resistance. Electrical resistance (at room temperature) of sintered composites is determined by using two probe methods.

Keywords: Industrial wastes, pond ash, Feldspar, XRD, phases, mullite, morphology, electrical resistance, and insulator

---

*Corresponding author: muktikanta2@gmail.com

9.1 Introduction

Electricity is backbone of modern world. Mostly, machineries require electricity for manufacturing process. Therefore, electricity is a driving force for global economy growth. Coal is the main source for electricity production. In thermal power stations, tons of fly ash/bottom ash/pond ash is generated. Thermal power stations always experience disposal problem of these materials. In addition, they also pollute our environment. In the present scenario, Fly ash is used for making value-added products. With present technological advancements, it has become possible to utilize fly ash for making value-added products. This material is used in construction industry, structural fill, making pavement, soil reclamation, soil ameliorant, an additive in anaerobic digestion, composting, zeolite synthesis, metal recovery, low-cost adsorbent for various gaseous, etc. [1–10].

When pulverized coal is fed to combustion chamber of boiler, instantly it ignites and generate molten residue. After cooling, left over hard mass becomes ash. Coarser part of hard mass is called bottom ash. Finer suspended particle is separated by electrostatic precipitors (ESPs).

Now-a-days, ash is dumped into pond. Therefore, it is called pond ash. Pond ash is being generated in an alarming rate. The generation of pond ash is posing a threat to our environment and the management is now experiencing problem to dispose it and these have become the thrust area for engineering research [11].

Physical and chemical characteristics of thermal power plants ashes depend on source of coal and burning process in power plant [12]. Major elements of ash are silicon, calcium, aluminum, iron, magnesium, sulfur, carbon [13]. Looking to the compositions, efforts are made to develop new routes for their utilization. One such way is to use the wastes for making ceramic materials. For the development process, large mass raw materials and high temperatures are required to make it possible.

For improving insulating properties of raw pond ash, other mineral is need to be added for making composite material. A few published articles related to thermal power plant ash based composite, particularly, pond ash based composite are available. Therefore, there is large scope to explore the possibility of using pond ash for making insulator. For Insulator production, DC resistivity, dielectric constant, and dielectric loss are main considerations [14].

In 1850, electrical insulators are employed for construction of electrical air lines. These electrical insulators are made from ceramic materials [15] Ceramic materials are having high mechanical strength, high dielectric strength, and good corrosion resistance [16, 17]. Also, Ceramic materials are used in various purposes such as refractories [18], cutting tools [19], thermal insulations [20], etc. For ceramic material production, high pressure and high temperature are generally, required. Performances of ceramic materials are evaluated by presence of crystallized phase(s). Such phases are achieved by controlling of sintering process parameters. Mullite phase enhances resistance of the ceramic material [21, 22]. Formation of mullite phase in the ceramic materials is a key issue and is possible at sintering temperature between 1150 to 1300 °C [23]. Work is going on to improve the dielectric and electrical insulation properties of industrial wastes [24–27].

One of the minerals, Feldspars plays key role because of their availability [28]. Feldspars are mainly rock-forming minerals present in magmatic, metamorphic, and sedimentary rocks. Therefore, conductivity of feldspars has great influence on electrical conductivity of rocks in present Earth's crust. Alkali feldspar helps to add electrical properties of whole feldspar system. Various reports are available on electrical properties of feldspar [29–32]. Transport phenomenon of feldspar is due to migration of Na^+, K^+ and Ca^{2+} cations. But, this phenomenon is yet to be established unclear. It is believed that movement of alkali cations is controlled by a vacancy mechanism or by an interstitial mechanism. However, this concept is still a controversial issue. Furthermore, electrical conductivity of albite phase has been reported [33]. Therefore, clarification of dynamics of charge transport in alkali feldspar at high temperature is important.

In the present time, cost-effective insulator is needed in the electrical sector. Current demands motivate researchers to develop the insulators having best properties to meet our present requirements [34]. Insulators have great importance in our day-to-day life. It makes our life easy, harmless and shock free [35–37].

Present investigation of the work is to prepare sintered PA/CC/PY hybrid materials through sintering at 1200 °C via solid state route. Morphology and elemental analyses are done by FESEM with EDS of the raw and sintered materials. Presences of various phases of the materials have been investigated by X-ray diffraction technique. Occurrence of chemical groups is identified by FTIR analyses of the samples. Electrical resistivity value is estimated at room temperatures by two-probe method.

9.2 Experimental Details

9.2.1 Materials and Chemicals

The main component (i.e., Pond ash) is received from National Aluminium Company (NALCO), Odisha, India. Feldspar and dextrin $(C_6H_{10}O_5)_n \cdot xH_2O$) are procured from Merck India. Few drops of water (*i.e.*, 6%) is added to make the composite pasty. Different compositions of pond ash and feldspar are indicated in Table 9.1. Optical images of pond ash, dextrin, and feldspar are shown in Figure 9.1.

9.3 Preparation of PA/FD Sintered Materials

PA/FD composite materials are attainable through solid state route followed by sintering process [40]. Different proportions of PA/k-FD composites are displayed in Table 9.2. Steps are followed as follows;

Step-1 Processing of raw materials (i.e., pond ash)
At first, as received pond ash is grinded in a ball mill for 5 h. It is sieved through 240 meshes. Sieved pond ash is dried in a heating oven at 120 °C (2 h). Purpose of heating sieved pond ash is to remove the existence of moisture. This pond ash is called processed pond ash.

Step-2 Preparation of green pellet
In this step, two types of mixtures are made. For preparation of composite, different proportions of components are used and are indicated in Table 9.2. Appropriate amount of processed pond ash is kept in a motar with pastel. A requisite amount of feldspar is added to the motar containing pond ash (Mixture-1). It is grinded for 2 h in order to get a uniform mixture. A small amount of dextrin (i.e., 0.05 %) is further added to mixture-1 and is grinded for 1 h (Mixture-2). Small amount of water is added to make the mixture pasty. Such paste samples are known as green samples. Samples are ready for preparation of pellets.

Pellets are made using pelletizer and universal testing machine (UTM). Components of pelletizer are properly cleaned, washed, and dried before assembling them. Appropriate amount of samples is kept in the pelletizer. Compaction pressure is maintained to the sample containing pelletizer. 10 MPa compaction pressure is applied for five minutes. After that, pelletizer is removed from moulding system and pellets from pelletizer are demoulded. Pellets are collected. Such pellet is ready for sintering.

Table 9.1 Constituents with percentage (%) of as-processed pond ash and feldspar.

S. ID	Compositions (%)													Ref.
	Fe_2O_3	Al_2O_3	SiO_2	P_2O_5	Cr_2O_3	ZnO	MgO	C	CaO	TiO_2	MnO	LOI		
PA	3.85	28.3	62.8	0.32	0.04	0.027	0.049	1.15	0.7	1.84	0.03	0.5	[38]
FD	---	18.18	64.57	---	0.22	---	---	---	---	0.01	---	---	[39]

Note: S. ID=Sample ID, PA=Pond ash, Feldspar, T=Trace.

164 High Electrical Resistance Ceramics

Figure 9.1 Optical image of pond ash (a), dextrin (b), and k-feldspar (c).

Table 9.2 Compositions (%) of pond ash/k-FD materials preparation.

S. no.	Individual composition name	
	Pond ash	k-Feldspar (k-FD)
1	50	50
2	60	40
3	70	30
4	80	20
5	90	10

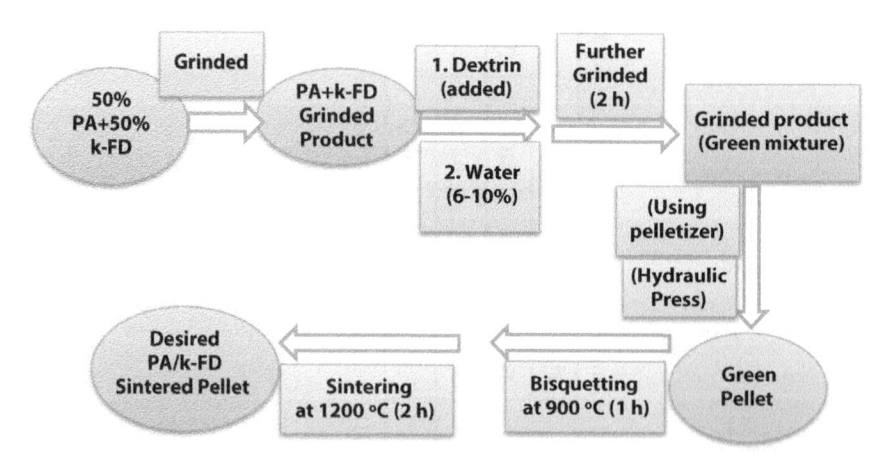

Figure 9.2 (Scheme 2) Flow chart for preparation of sintered pond ash/k-Feldspar (50:50) based materials.

Step-3 Sintering of green pellets
Green pellet samples are kept in muffle furnace. They are bisquetted at 900 °C (for 1 h) to remove intacted water and binder. Bisquetted pellet(s) are then, sintered at 1200 °C (for 2 h). Average dimensions (i.e, thickness and diameter) of sintered pellets are found to be 0.5 cm and 3 cm, respectively. The entire process is described schematically in Figure 9.2. Sintered pellet(s) is ready for different characterization process.

All other sets shown in Table 9.2 have followed similar treatment as described earlier.

9.4 Test Methods

XRD pattern is obtained with the help Phillips PW-1710 advanced wide angle X-ray diffractometer (Phillips PW-1729 X-ray generator instrument). CuKα radiation is used for the diffraction process. Diffraction patterns have been taken from pellet and untreated pond ash. For this measurement, samples are loaded in a quartz sample holder. Samples are scanned between 10° to 70° with scanning speed of 2°/min.

Optical image is used for measurement of hardness by Nano indentation (AntonPaar-NHT, Switzerland).

Field emission scanning electron microscopic (FESEM) tests are done. Carl Zeiss Supra 40 is used to study the surface topology of samples. Gold coating is done through sputtering technique. FESEM is operated at 30 kV. Elemental analyses are done by EDS, which is attached with FESEM machine.

5022/22, TecnaiG220 S-Twin, Czech Republic Transmission Electron Microscope (TEM) is used to investigate images and SAD pattern of raw pond ash (PA) powder and pond ash/k-Feldspar (60:40) composite. TEM specimen (raw PA) is prepared by drop casting technique. TEM specimen of pond ash/k-Feldspar (60:40) composite is prepared by microtone technique (LEICA Microsystem, GmBH, A-1170). Then, specimens are transferred to carbon coated Cu TEM grids.

FTIR spectra of samples (i.e., pond ash (PA), k-feldspar, and sintered PA based composite) are recoded in absorbance mode utilizing Thermo Nicolt Nexus 870 spectrophotometer (range 400-4000cm^{-1}). For these analyses, KBr is mixed with test samples and pellets are prepared using compression molding technique. Before analyses, background spectrum is recoded.

Resistivity values of samples are estimated by electrical measurement using following relation [40].

$$\rho = \frac{R \times A}{l} \qquad (9.1)$$

Here, A is the cross-sectional area of electrode i.e., $\frac{\pi}{4} D^2 \text{(centimeter)}^2$;

R is the resistance of prepared samples (MΩ);
l is average thickness of the sample (centimeter);
ρ is the resistivity of the samples in Ω.cm.

9.5 Results and Discussion

Table 9.3 describes resistance values of sintered pond ash and sintered pond ash/k-feldspar composite prepared in different proportions. Sintering temperature is used 1200 °C. From resistance data, sintered pond ash/k-feldspar (at 1200 °C) having 60% pond ash and 40% k-feldspar composition shows the highest resistance value (resistivity value) which is 18.3 GΩ (258.66135 × 10^9 Ohm.cm), whereas sintered pond ash (1200 °C) shows the lowest resistance value (resistivity value) which is 0.58 GΩ (8.19801 × 10^9 Ohm.cm). This is attributed to mullite formation at higher sintering temperatures [41].

Figure 9.3 displays XRD pattern of Pond ash (a), k-Feldspar (b), Pond ash/k-FD(50:50, c), Pond Ash/k-FD(60:40, d), Pond Ash/k-FD(70:30, e). These materials are bisquited followed by sintering at 1200 °C. Table 9.4 specifies crystal system and crystallographic parameters (i.e., unit cell dimension, crystal angle, space group, space group number, and density) of identified phases.

Various phases identified are alumina, silica, and mullite (Figure 9.3). Most prominent peak is obtained for silica phase. Other two dominating peaks are mullite and alumina (Figure 9.3) [41].

Observation is made from Figure 9.3 that all samples show presence of mullite phase. However, there is a variation in quantity of mullite phase for different samples. Extent of mullite phase formation depends on conversion of silica and alumina due to sintering reaction occurring between pond ash and k-feldspar components. Conversion of mullite phase by reaction between Alumina and silica component of raw materials is evidenced from the binary phase diagram (SiO_2 and Al_2O_3). Mullite phase is dominated at 1200 °C [42].

Optical image of pond ash/k-Feldspar (60:40) composite sintered at 1200 °C is displayed in Figure 9.4. It can be noticed from Figure 9.4 that

Table 9.3 Resistance values of pond ash (sintered) and sintered pond ash/k-Feldspar composites [radius=1.5 cm and thickness =0.5 cm].

S. no.	Sample ID	Sintering temperature (°C)	Resistance value	Resistivity (Ohm.cm)
1	Sintered Pond ash	1200	0.58 GΩ	8.19801×10^9
2	Pond ash (50%)/ k-Feldspar (50%)	1200	6.2 GΩ	87.6339×10^9
3	Pond ash (60%)/ k-Feldspar (40%)	1200	18.3 GΩ	258.66135×10^9
4	Pond ash (70%)/ k-Feldspar (30%)	1200	15.4 GΩ	217.6713×10^9
5	Pond ash (80%)/ k-Feldspar	1200	12..8 GΩ	180.9216×10^9
6	Pond ash (90%)/ k-Feldspar (10%)	1200	1.26 GΩ	17.80947×10^9

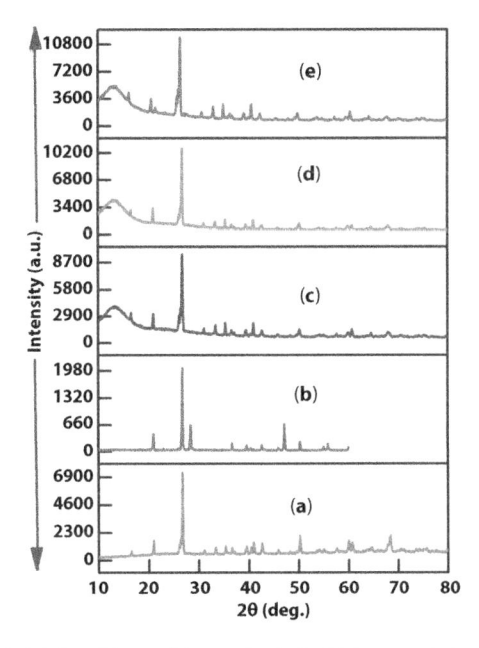

Figure 9.3 Pond ash (a), k-Feldspar (b), Pond ash /k-FD(50:50, c), Pond ash/k-FD(60:40, d), Pond ash/k-FD(70:30, e).

Table 9.4 Structural information of pond ash/k-Feldspar based samples.

Crystal ID	Mineral name	Crystal system	Space group	Space group number	Cell dimension	Crystal angle	Calculated density (g/cm^3)
As-received PA	Quartz SiO_2	Hexagonal	P3221	154	a=b≠c	α=β=90°, γ=120°	2.65
PA 1200 °C	Mullite $Al_2(Al_{2.5}Si_{1.5})O_{9.75}$	Orthorhombic	Pbam	55	a≠b≠c	α=β=γ=90°	3.17
PA 1200 °C	Quartz SiO_2	Cubic	Fd-3m	227	a=b=c	α=β=γ=90°	2.21
PA+k-FD (60%+40%) 1200 °C	Mullite $Al_{4.52}Si_{1.48}O_{9.74}$	Orthorhombic	Pbam	55	a≠b≠c	α=β=γ=90°	3.17
PA+k-FD (60%+40%)	Quartz SiO_2	Hexagonal	P3221	154	a=b≠c	α=β=90°, γ=120°	2.64
PA+k-FD (70%+30%) 1200 °C	Mullite $Al(Al_{1.272}Si_{0.728}O_{4.864})$	Orthorhombic	Pbam	55	a≠b≠c	α=β=γ=90°	3.16

(Continued)

Table 9.4 Structural information of pond ash/k-Feldspar based samples. (*Continued*)

Crystal ID	Mineral name	Crystal system	Space group	Space group number	Cell dimension	Crystal angle	Calculated density (g/cm^3)
PA+ k-FD (70%+30%) 1200 °C	Quartz SiO$_2$	Hexagonal	P3221	154	a=b≠c	α=β=90°, γ=120°	2.63
PA+ k-FD (80%+20%) 1200 °C	Silicon dioxide	Hexagonal	P3221	154	a=b≠c	α=β=90°, γ=120°	15.89
PA+ k-FD (80%+20%) 1200 °C	Mullite Al(Al$_{.83}$Si$_{1.08}$O$_{4.85}$)	Orthorhombic	Pbam	55	a≠b≠c	α=β=γ=90°	3.10

*Note: PA-Pond Ash, k-FD-potassium feldspar, SiO$_2$-Silicon dioxide.

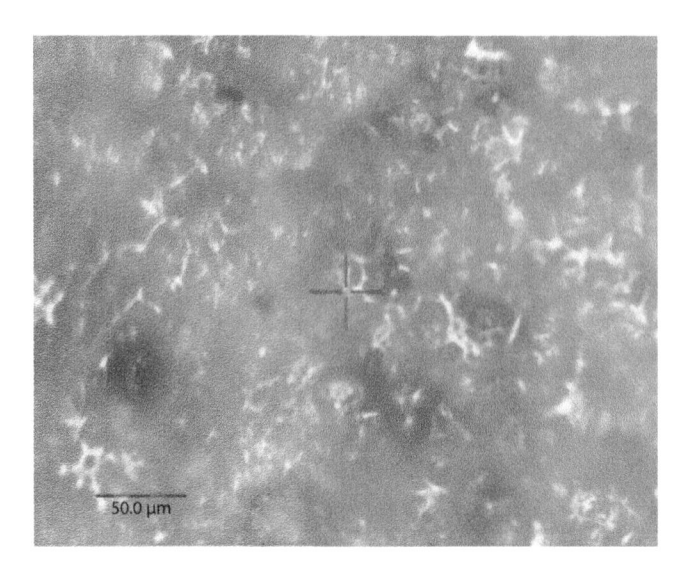

Figure 9.4 Optical micrograph of pond ash/k-FD composite.

white patches have spreaded in the matrix. There are irregular shaped particles having different textures, which are dispersed in the matrix. Optical micrograph shows poor contrast and so, FESEM study is carried out to understand these features in details.

Figure 9.5 show FESEM images of raw pond ash, k-feldspar, and pond ash/k-feldspar (60:40) composite sintered at 1200 °C (lower and higher magnification). Figure 9.5a displays FESEM image of raw pond ash. The texture is mainly organized by irregular elliptical particles having major and minor axis ratio 2:1 (approx.).

Figure 9.5b indicates k-feldspar particles. It is flecky crystalline structure. Figure 9.5c designates FESEM image of pond ash/k-feldspar (60:40) composite sintered at 1200 °C (magnification 5kX). From FESEM image, it is observed that pond ash/k-feldspar composite shows non-uniform compacted structure with small pores. This is formed due to the interaction of k-feldspar and pond ash particle, which is expected to be advantageous due to increasing crystallinity of the sintered composite. Crystalline structure is co-related with XRD pattern (Figure 9.3) which shows presence of mullite phase.

Figure 9.5d shows FESEM image of pond ash/k-feldspar (60:40) composite sintered at 1200 °C at a higher magnification of 20kX. FESEM image of pond ash/k-feldspar (60:40) composite sintered at 1200 °C shows non-uniform randomly oriented small fibers with compacted structure. A few small pores are observed in the image. This is formed due to the interaction of k-feldspar with PA spherical particles..

HIGH RESISTANCE SINTERED POND ASH/K-FELDSPAR CERAMICS 171

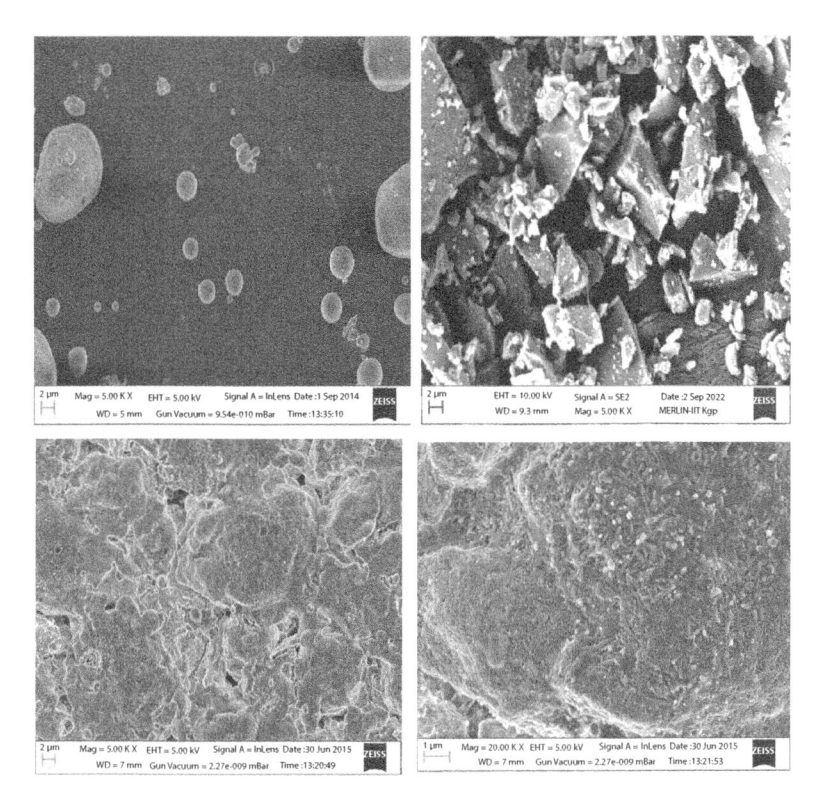

Figure 9.5 FESEM image of pure PA (a), Pure k-feldspar (b), PA/k-feldspar composite with lower magnification (c), and PA/k-feldspar composite with higher magnification (d).

Figure 9.6 shows EDS analyses of pond ash/k-feldspar (60:40) composite sintered at 1200 °C. It is obtained during FESEM studies. From EDS profile, elements like O, Al, Si, Ti, Fe, and K are commonly present in the sample. Ti and Fe elements are present in trace amount in the sample. EDS results are well-supported by XRD results.

TEM image of unsintered pond ash is shown in Figure 9.7a. There are ellipsoidal particle with varying aspect ratios embedded in pond ash particles. Some of them are found to be cylindrically in nature. Similar impression is made by observing FESEM micrograph of the same material (Figure 9.5). Diffraction pattern is obtained from a spot which shows bright spots, arranged in different circumferences around the centre indicating crystallinity of the phases present in TEM micrograph (Figure 9.7a).

Figure 9.8a shows TEM of pond ash/k-feldspar (60:40) sintered composite. Some unreacted particles of pond ash are dispersed in the matrix, which are crystalline in nature. Some area shows fused particles, which is due to reaction occurring between pond ash and k-feldspar at the sintering

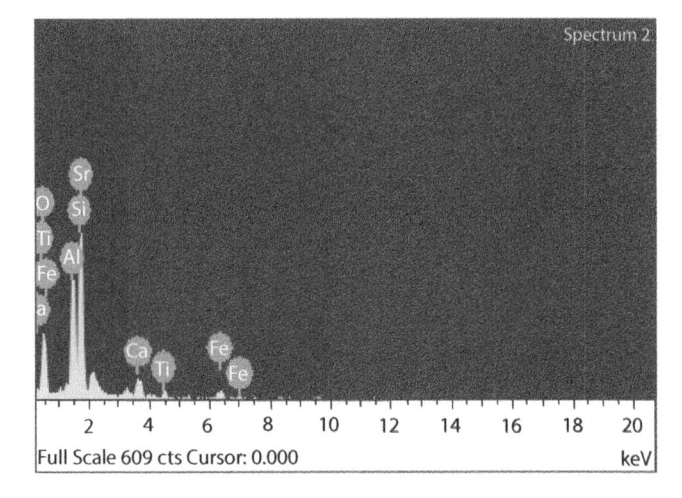

Figure 9.6 EDS spectrum of pond ash/k-feldspar (60:40) composite sintered at 1200 °C.

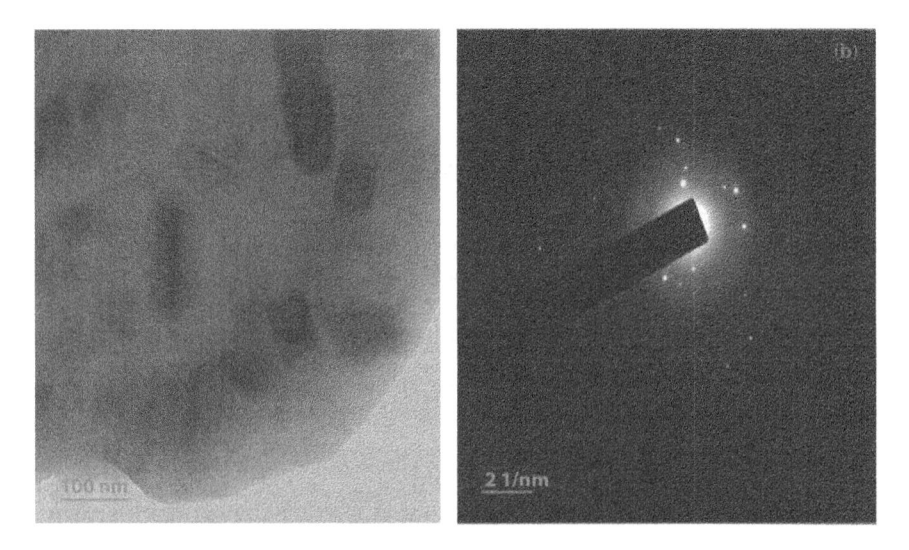

Figure 9.7 HRTEM image (a) and SAD pattern (b) of pond ash.

temperature. Diffraction pattern of the sintered composite is snown in Figure 9.8b. Here bright spots are concentrated with lesser spots present around the circumference of circle. Crystallinity of sintered products is maintained in the composite product.

Figure 9.9a displays FTIR spectrum of k-feldspar. Band occurring at 2919.50 cm^{-1} corresponds to O-H bonds and H bending vibrations of H-O-H of interlayer adsorbed H_2O molecule band. Band at 1449.40 cm^{-1}

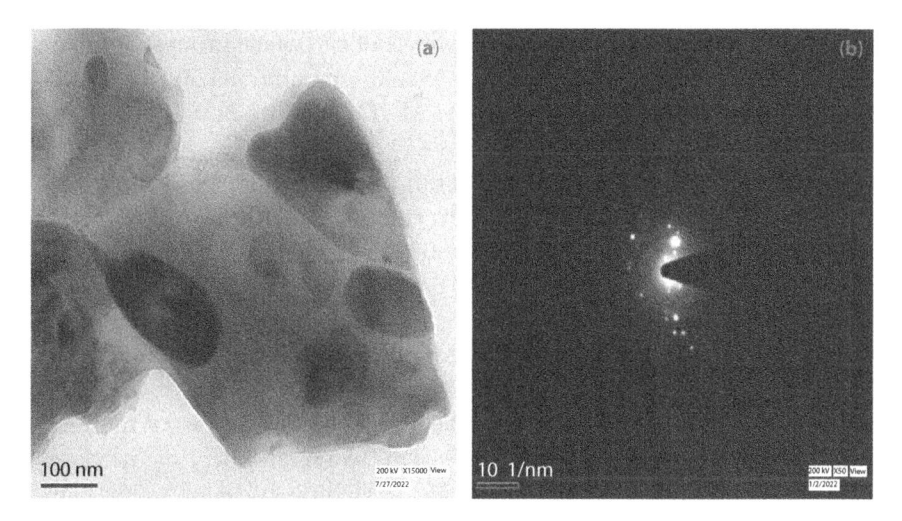

High Resistance Sintered Pond Ash/k-Feldspar Ceramics 173

Figure 9.8 TEM image (a) and SAD pattern (b) of pond ash/k-Feldspar (60/40) composite.

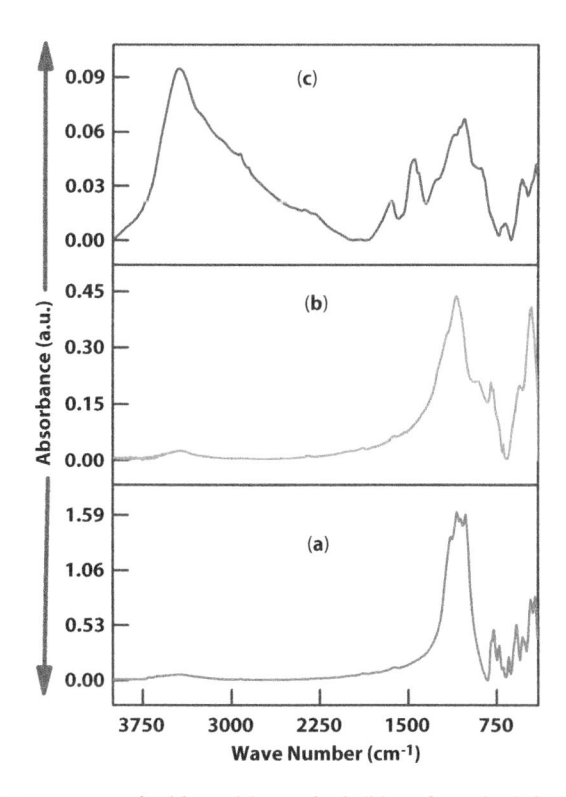

Figure 9.9 FTIR spectrum of Feldspar (a), pond ash (b), and pond ash/k-Feldspar (60:40) sintered composite.

agrees to C=O bond. It signifies occurrence of carbonate groups. Si-O and O-Si-O bands are also established at 989.30 cm^{-1} and 534.50 cm^{-1}. These two bands indicate the silicate groups. Al^{3+}·O^{2-} bands is found 805.5 cm^{-1} [43–49]. Fe-O band is also observed at 440 cm^{-1} [43–49].

Figure 9.9b indicates FTIR spectrum of raw pond ash. Occurrences of various band of pond ash are in agreement with different components pond-ash. Bands at 600, 1098, and 1608 cm^{-1} match with Si-O-Al band, Si-O-Si asymmetric band and HO-H bending band respectively [43–49].

Figure 9.9c exhibits FTIR spectrum of Pond Ash/k-Feldspar (60:40) composite. Occurrence of bands are found at different wave numbers i.e., 3441, 2918, 1823, 1638, 1117 and 907 cm^{-1} (Figure 9.9c). Bands are established with stretching vibrations of O-H bonds (3441 cm^{-1} wave number) and H-O-H bending vibrations (1638 cm^{-1} wave number) of interlayer adsorbed H$_2$O molecule. The hydroxyl-stretching band of water plays an important role and peak shift of the FTIR spectra is significant. Band (at 1117 cm^{-1}) is attributed to Si-O group. It signifies presence of silicate groups. Band (near 907 cm^{-1}) is indicated to Al^{3+}O^{2-} group [43–49].

9.6 Conclusions

Results of the work have indicated that sintered pond ash (60%)/k-feldspar (40%) composite have yielded material which can substitute conventional insulators. Optimization of the process parameter has enabled an insulator having specific resistance 18.3 MΩ (258.66135 × 10^9 Ohm.cm) sintered to be produced.

XRD pattern has revealed mineralogical phases present in both raw pond ash and sintered composite. Quartz is the dominating phase present in raw pond ash and sintered composite. By sintering at 1200 °C, formation of the mullite phase has occurred. Mullite phase is a microreinforcement phase, which adds strength and resistivity to the material structure.

FESEM micrographs have indicated that with increasing sintering temperature the powder particles of pond ash and flaky structure of k-feldspar are becoming more compact with increasing temperature due to agglomeration of particles by incipient fusion of reactants.

FTIR spectra have shown noticeable variance. The corresponding groups of the peaks are designated formation of new bonds within the phases.

Acknowledgements

First author would like to thank Prof. Munesh Chandra Adhikary, PG Council Chairman, Fokir Mohan University, for his invaluable guidance, advice, and constant inspiration throughout the entire program. First author thanks Mr. Mukteswar Mohapatra in Fokir Mohan University for his support. The authors convey their sincere thanks to GIET, University Gunupur, Rayagada, Odisha, India for providing Lab facilities to do the research work. The authors would like to thank the CRF, IIT Kharagpur for providing their testing facilities.

References

1. S. J. Ahn, and D. Graczyk, (2012), Understanding Energy Challenges in India. Policies, Players and Issues, International Energy Agency (IEA).
2. S. Kumar, G. Mucsi, F. Kristály, and P. Pekker, (2017), Mechanical Activation of Fly Ash and its Influence on Micro and Nano-Structural Behaviour of Resulting Geopolymers, *Advanced Powder Technology*, Vol. 28, pp. 805-813.
3. L. C. Ram and R. E. Masto, (2014), Fly Ash for Soil Amelioration: A Review on the Influence of Ash Blending with Inorganic and Organic Amendments, *Earth Science Review*, Vol 128, pp. 52-74.
4. S. A. Haldive and A. R. Kambekar, (2013), Experimental Study on Combined Effect of Fly Ash and Pond Ash on Strength and Durability of Concrete, *International Journal of Scientific and Engineering Research*, Vol. 4, pp. 81-86.
5. V. C. Pandey and N. Singh, (2010), Impact of Fly Ash Incorporation in Soil Systems, Agriculture, *Ecosystems & Environment*, Vol. 136, pp. 16-27.
6. M. L. D. Jayaranjan, E. D. Van Hullebusch, and A. P. Annachhatre, (2014), Reuse Options for Coal Fired Power Plant Bottom Ash and Fly Ash, *Reviews in Environmental Science and Bio/Technology*, Vol. 13, pp. 467-486.
7. MOEF, Gazette notification for Ministry of Environment and Forests, No. 563. New Delhi: Ministry of Environment and Forests, 14 September 1999.
8. Central Electricity Authority (CEA) Report 2018-2019. Available online at: http://cea.nic.in/reports/others/thermal/tcd/flyash_201819.pdf (accessed on 05-04-2020).
9. M. Ahmaruzzaman, (2010), A Review on the Utilization of Fly Ash, *Progress in Energy and Combustion Science*, Vol. 36, pp. 327-363.
10. Z. T. Yao, X. S. Ji, P. K., Sarker, J. H. Tang, L. Q. Ge, M. S. Xia, and Y. Q. Xi, (2015), A Comprehensive Review on the Applications of Coal Fly Ash, *Earth Science Revew*, Vol. 141, pp. 105-121.

11. P. Ghosh and S. Goel, (2014), Physical and Chemical Characterization of Pond Ash, *International Journal of Environmental Research and Development*, Vol. 4, pp. 129-134.
12. L. Benco, D. Tunega, J. Hafner and H. Lischka, (2001), Ab initio Density Functional Theory Applied to the Structure and Proton Dynamics of Clays, *Chemistry Physics Letter*, Vol. 333, pp. 479-484.
13. J. C. Hower, C. L. Senior, E. M. Suuberg, R. H. Hurt, J. L. Wilcox, and E. S. Olson, (2010), Mercury Capture by Native Fly Ash Carbons in Coal-Fired Power Plants, *Progress in Energy and Combustion Science*, Vol. 36, pp. 510-529.
14. D. E. Yıldız and I. Dokme, (2011), Frequency and Gate Voltage Effects on the Dielectric Properties and Electrical Conductivity of $Al/SiO_2/p$-Si Metal-Insulator-Semiconductor Schottky Diodes, *Journal of Applied Physics*, Vol. 110, 014507.
15. W. V. Siemens, (1966), Inventor and Entrepreneur: Recollections of Werner von Siemens, London, England.
16. T. Tanaka and T. Imai, (2013), Advances in Nanodielectric Materials over the past 50 years, *IEEE Electrical Insulation Magazine*, Vol. 1, pp. 10-23.
17. T. Tanaka, (2005), Dielectric Nanocomposites with Insulating Properties, *IEEE Transactions on Dielectrics and Electrical Insulation*, Vol. 12, pp. 914-928.
18. D. Goski and M. Lambert, (2019), Engineering Resilience with Precast Monolithic Refractory Articles, *International Journal of Ceramic Engineering & Science*, Vol. 1, pp. 169-177.
19. L. Li and Y. Li, (2017), Development and Trend of Ceramic Cutting Tools from the Perspective of Mechanical Processing, *IOP Conf. Series: Earth and Environmental Science*, Vol. 94, pp.1-6.
20. M. Zhu, R. Ji, Z. Li, H. Wang, L. L. Liu, and Z. Zhang, (2016), Preparation of Glass Ceramic Foams for Thermal Insulation Applications from Coal Fly Ash and Waste Glass, *Construction and Building Materials*, Vol. 112, pp. 398-405.
21. S. B. Kondawar, A. D. Dahegaonkar, V. A. Tabhane, and D. V. Nandanwar, (2014), Thermal and Frequency Dependence Dielectric Properties of Conducting Polymer/Fly Ash Composites, *Advanced Materials Letters*, Vol. 5, 360.
22. J. Martín-Márquez, A. G. De la Torre, M. A. G. Aranda, J. Ma Rincón, M. Romero, (2009), Evolution with Temperature of Crystalline and Amorphous Phases in Porcelain Stoneware, *Journal of the American Ceramic Society*, Vol. 92, pp. 229-234.
23. H. E. Exner and E. Arzt, (1996), Sintering Processes. In Physical Metallurgy, Ed. R. W. Cahn and P. Haasen. 4th Ed. Elsevier Science, Amsterdam, Vol. 3, pp. 2628-2662.

24. A. Pattnaik, S. K. Bhuyan, S. K. Samal, A. Behera, and S. C. Mishra, (2016), Dielectric Properties of Epoxy Resin Fly Ash Composite, *IOP Conference Series: Materials Science and Engineering*, Vol. 115.
25. M. Mishra, A. P. Singh, and S. K. Dhawan, (2013), Utilization of Fly Ash-A Waste By-product of Coal for Shielding Application, *Journal of Environmental Nanotechnology*, Vol. 2, 74.
26. J. P. Dhal, and S. C. Mishra, (2013), Investigation of Dielectric Properties of a Novel Hybrid Polymer Composite using Industrial and Bio-waste, *Journal of Polymer Composites*, Vol. 1, 22.
27. K. Singh, T. Quazia, S. Upadhyaya, and P. Sakharkarb, (2005), Development of Low Permittivity Material using Fly Ash, *Indian Journal of Engineering & Materials Sciences*, 12, 345.
28. E. C. Nzenwa and A. D. Adebayo, (2019), Analysis of Insulators for Distribution and Transmission Networks, *American Journal of Engineering Research (AJER)*, Vol. 8, pp. 138-145.
29. C. L. Goldsmith, A. Malczcwski, J. J. Yao, S. Chen, J. Ehmk, and D. H. Hinzel, (1999), RFMEMS-Based Tunable Filters, *International Journal of RF and Microwave Computer-Aided Engineering*, Vol. 9, 362.
30. G. Subramanyam, F. V. Keuls and F. A. Miranda, (1998), Novel K-band Tunable Microstrip v-band Pass Filter using Thin Film HTS/Ferroelectric/Dielectric Multilayer Configuration, *IEEE Microwave Guided Wave Letter*, Vol. 8, 78.
31. A. Tombak, J. P. Maria, F. T. Ayguavives, Z. Jin, G. T. Stauf, A. I. Kingon, A. Mortazawi, (2003), Voltage-Controlled RF Filters Employing Thin Film Barium-Strontium Titanate Tunable Capacitors, *IEEE Transection Microwave Theory and Technology*, Vol. 51, 462.
32. M. K. Panigrahi, R. R. Dash, R. I. Ganguly, (2018), Development of Novel Constructional Material From Industrial Solid Waste as Geopolymer for Future Engineers, *IOP Conference Series: Materials Science and Engineering*, Vol. 410, pp.1-12,
33. M. K. Panigrahi, P. Kumar, B. Barik, D. Behera, S. K. Mohapatra, H. Jha, (2016), Frequency Dependency of Developed Dielectric Material from Fly Ash: An Industrial Waste, *20th National Conference on Nonferrous Minerals and Metals*, 8-9th July 2016; Eds. Rakesh Kumar, K.K.Sahu & Abhilash, pp.143-150.
34. T. R. Naik, R. Kumar, B. W. Ramme, and R. N. Kraus, (2010), Effect of High-Carbon Fly Ash on the Electrical Resistivity of Fly Ash Concrete Containing Carbon Fibers, Second International Conference on Sustainable Construction Materials and Technologies June 28th, Università Politecnica delle Marche, Ancona, Italy.

35. F.H. Wee, F. Malek, S. Sreekantan, A.U. Al-Amani, F. Ghanil, K.Y. You, Investigation of the Characteristics of Barium Strontium Titanate (BST) Dielectric Resonator Ceramic Loaded on Array Antennas, *Progress in Electromagnetics Research*, 121 (2011) 181-213.
36. G. Subramanyam, F.V. Keuls and F. A. Miranda, Novel K-band Tunable Microstrip v-band Pass Filter using Thin Film HTS/Ferroelectric/Dielectric Multilayer Configuration, *IEEE Microwave and Guided Wave Letters*, 8 (2) (1998) 78-80.
37. https://en.wikipedia.org/wiki/Insulator_(electricity).
38. H. Hu, H. Li, L. Dai, S. Shan, and C. Zhu, (2013), Electrical Conductivity of Alkali Feldspar Solid Solutions at High Temperatures and High Pressures, *Physics and Chemistry of Minerals*, Vol. 40, pp. 51-62.
39. Development of Novel Constructional Material From Industrial Solid Waste as Geopolymer for Future Engineers Muktikanta Panigrahi, , Radha Raman Dash, Ratan Indu Ganguly, in *IOP Conference Series: Materials Science and Engineering*, 2018, 410 (1), pp.1-12.
40. M.K. Panigrahi, (2021), Investigation of Structural, Morphological, Resistivity of Novel Electrical Insulator: Industrial Wastes, *Bulletin of Scientific Research*, Vol. 3, pp. 51-58.
41. M. K Panigrahi, P. K. Rana, A. K. Pradhan, P. K. Rout, A. K. Samal, S. Gupta and Mv B. Kumar, (2015), Production of Geopolymer based Construction Material from Pond Ash: An Industrial Waste, *19th International Conference Non-Ferrous Metal*-2015 (9,10th July), pp. 190-200.
42. R. Roy, D. Das, and P. K. Rout, (2022), A Review of Advanced Mullite Ceramics, *Engineered Science*, Vol. 18, pp. 20-30.
43. A. Jose, M. R. Nivitha, M. Krishnan, and R. G. Robinson, (2020), Characterization of cement stabilized pond ash using FTIR spectroscopy, *Construction and Building Materials*, Vol. 263, 120136.
44. Y. Liu, F. Zeng, B. Sun, P. Jia, and I. T. Graham, (2019), Structural Characterizations of Aluminosilicates in Two Types of Fly Ash Samples from Shanxi Province, North China, *Minerals*, Vol. 9, pp. 358-16.
46. P. Rożek, M. Król, and W. Mozgwa, (2018), Spectroscopic studies of fy ash-based geopolymers, *Spectrochim Acta - Part A Mol Biomol Spectrosc* Vol. 198, pp. 283-289.
47. S. Kumar, F. Kristály, and G. Mucsi, (2015), Geopolymerisation behaviour of size fractioned fly ash, *Adv Powder Technol*, Vol. 26, pp. 24-30, https://doi.org/10.1016/j.apt.2014.09.001.

48. Z. Kledyński, A. Machowska, B. Pacewska & I. Wilińska, (2017), Investigation of hydration products of fly ash-slag pastes, *Journal of Thermal Analysis and Calorimetry*, Vol. 130, pp. 351-363, https://doi.org/10.1007/s10973-017-6233-4.
49. D. Bondar and R. Vinai, (2022), Chemical and Microstructural Properties of Fly Ash and Fly Ash/Slag Activated by Waste Glass-Derived Sodium Silicate, *Crystals*, Vol. 12, pp. 913-12, https:// doi.org/10.3390/cryst12070913.

10

Applications, Challenges and Opportunities of Industrial Waste Resources Ceramics

Muktikanta Panigrahi[1]*, Ratan Indu Ganguly[2] and Radha Raman Dash[3]

[1]*Department of Materials Science, Maharaja Sriram Chandra Bhanja Deo University, Balasore, Odisha, India*
[2]*Department of Metallurgical Engineering, National Institute of Technology, Raurkela, Odisha, India*
[3]*CSIR-National Metallurgical Laboratory, Jamshedpur, Jharkhand, India*

Abstract

Wastes from different manufacturing processes and energy generation units possess different health issues. Instead of land-filling by them, they can be recycled or reused by conver ting them into marketable value-added products. In addition, there will be gain in the economy of the regions where they are produced. From eco-friendly propensity in the last two decades, there are researches which show encouraging outcomes. It is observed that waste material can be used as raw materials for producing useful product. Finding show possibilities of using waste material. They are demonstrated the possibility to use alternative ingredients in the place of conventional raw materials (e.g., most common ternary clay-quartz-feldspar system) for the fabrication of ceramics. Researchers are trying to incorporate the wastes and industrial by-products like fly ash (FA), rice husk ash (RHA), blast furnace slag (BFS), sludge, glass waste, polished tile waste, eggshell and others for making different ceramics. Thus, conventional raw materials can be dispensed with by use of waste products as mentioned above. Present review is aimed to provide an update overview of waste derived ceramics which include refractories, glasses, white wares, oxide and non-oxide ceramics. Investigation reveals that ceramic industries have huge potential for utilization of wastes and substitution of natural raw materials. Converting wastes to value-added ceramics will not only solve disposal problems but also conserves natural resources.

Corresponding author: muktikanta2@gmail.com

Muktikanta Panigrahi, Ratan Indu Ganguly and Radha Raman Dash. *High Electrical Resistance Ceramics: Thermal Power Plants Waste Resources*, (181–198) © 2023 Scrivener Publishing LLC

Keywords: Industrial wastes, ceramic products, applications, opportunities, challenges

10.1 Introduction

Concrete is the world's largest amount of man-made materials. Presently, it is mostly used for infrastructure construction. By utilizing waste materials as resource materials, it is possible to reduce consumption of sand and gravels for construction industry. Consumption of natural resources such as sand and gravel aggregate will rapidly increase. It is estimated that concrete industries consume around 5 billion tons of natural aggregate such as sand and gravel as chief raw materials. As a result, deposits of these raw materials and store house like mountains are detecting. Consequently, the earth is losing ecological balance. Destruction of mountain creates ecological disbalance. It will also affect mountain landscape and green vegetation. Digging of sand is causing many problems such as soil erosion or river diversions, etc. In addition, many countries and regions have shortage of gravel sand, and concrete aggregate.

To address these problems, people have begun to seek new aggregate resources. Some success have been achieved to some extent.

In this quest, people are using waste materials as resource material for developing infrastructures [1–3].

During production, transportation, sale, storage and use of all kinds of ceramic products are get damaged. Thus, damaged materials get accumulated. In some places, aging and other factors get accumulated and they can be called as waste ceramics.

Literature survey has shown [4] that 30% of the world's ceramic industry products are industrial waste. Ceramic industry is being a traditional industry in China. Therefore, amount of waste is large. But literature survey indicates that there is no way for effective recycling of waste ceramic.

In order to tackle challenges in keeping with pace, ceramic industries and construction industry are trying to develop new materials to replace traditional raw materials. It is thought that effective utilization of ceramic waste may be one solution. The pollution problem is minimized and creates new opportunities for entrepreneurs to produce raw materials for civil and construction engineers. Reusing ceramic waste powder in building material production is not only efficient to use resources, but also reduces pollutant emissions. This will create conducive and sustainable development of society and nature. It is believed that mixing ready-mixed concrete products with ceramic powder waste to some extent can be considered environmental friendly concrete. H. Mizuguchi [5], X. Wang [6] has opined that environmental protection concrete will reduce environmental load and will improve ecological environment, in harmony with nature.

At present, not much research data is available from domestic scholars. Therefore, it is essential to investigate reusing of the materials. The gap of current technology exists between our country and the international advanced countries, mainly on utilization of waste products.

10.2 Different Ways of Utilization of Waste

In the present scenario, scientific community has put their effort for utilizing wastes, which is produced in different sectors. The present study has attempted a review on engineering applications of waste based ceramics products (with reference) with porous insulation refractory, dense refractories, ceramic tiles, waste glass powders, glass, glass-ceramics, Mullite, etc.

10.2.1 Porous Insulation Refractory

Generally, refractories are used for two purposes

1. Ceramic vessels are used as liner to protect metals/alloys. Corrosion and erosion is caused due to interaction between metal surface and hot flue gases, molten salts, liquid metals and slags.
2. To maintain inside required temperature of vessel (insulate). For first one, dense refractories with high refractoriness are used, because it is directly contacted with furnace or kiln environments. Low thermal conductivity (σ), moderate refractoriness, porous and lightweight refractories are generally used for insulation of furnaces. Researchers are mostly trying to incorporate wastes in insulation refractories. Present users are trying to use wastes in the composition of insulation refractories [7][a,b,c,d,e].

10.2.2 Dense Refractory

Some researchers have also attempted to introduce to utilize the waste ingredients for fabrication of high-temperature applicable dense working refractories. Khalil *et al.* (2018) [8] have synthesized the high-temperature refractory by using petroleum waste sludge (0 to100 wt.%) mixed with natural raw bauxite (100 to 0 wt.%). Mixture is fired at different temperatures restricted up to 1600°C. Petroleum sludge (oil processing waste) contains barium oxide (39.08 wt.%), silicon oxide (28.02 wt.%) and sulfur trioxide (21.9 wt.%) as main components. 40 wt.% of sludge and 60 wt.% of bauxite is considered as optimum composition, which have given best physico-mechanical properties. They meet requirements of international standards of refractories.

10.2.3 Ceramic Tiles

Tiles, the most popular and rapidly growing ceramics, are used in construction and building activities. Rapid urbanization, modernization, and renovation of older buildings are creating great demand for these ceramic materials. Therefore, government policies are giving priorities to infrastructure development. Global tiles market worth has projected at US$ 70.9 billion in 2018. Compound annual growth rate (CAGR) during 2011–2018 is estimated to be ~9.1%. Projected market revenue will be reaching above US$ 107.2 billion by 2024. It is expected that CAGR will rise by ~7.2% in the period 2019–2024 [9]. Different types of clay, silica, feldspar, zircon sand, alumina, and other natural resources are the key ingredients for the manufacturing of nearly all kinds of ceramic tiles. Huge consumption of naturally occurring minerals is causing several environmental issues. Therefore, environment-friendly substitution of natural ingredients is essential in forthcoming years. Some wastes and industrial by-products have been found as sustainable replacement of virgin raw materials in tiles. Certain key components like the amount of wastes, replacement of minerals (like clay, feldspar, and quartz), firing temperature, and categories (like porcelain, floor, wall, and glazed tiles) of waste-containing tiles are pointed in Table 10.1 [10]. Use of fly ash is rapidly extended for preparing of ceramic tiles. Chandra *et al.* (2008) [11] have prepared low-temperature firing wall tiles by addition of low alkali pyrophyllite and sodium hexameta phosphate (SHMP) with FA and fired in the temperature range 950 to1050°C.

Iron and Steel plants generate huge amount solid waste, i.e., slag in blast furnace as well as steel melting process. This slag usually contains SiO_2, Al_2O_3, CaO, MgO as major quantity and FeO, MnO_2, and TiO_2 as minor quantity [12]. Some companies are quenching blast Furnance slag at a very rapid rate to avoid any crystalline phase. This kind of slag is suitable for cement. However, 30 to 40% slags are mixed with clays and they can be used as ingredient for making tiles. They are ground to a very fine size and mixed with optimum amount of water and other chemicals. After sintering

Table 10.1 Name of waste used in the composition of tiles [10].

Name of wastes	Replacement of minerals	Atomic weight (%) of waste	Firing temperature (°C)	Types of tiles
Polished tiles waste	Proportionally replaced raw materials	50	1120	P. tiles
Fly ash	---	50	950	W. tiles
Fly ash-tincal waste	Feldspar	10, 5	1020	W. tiles
High Alumina Fly ash	Feldspar, Quartz	60	1200	---
High Alumina Fly ash	---	70	1300	---
Fly Ash	Clay, Feldspar, Quartz	100	1300	---
Cyclone dust, Filter dust (separately added)	---	75	1190	F. tiles
Sanitryware waste	Pegmatite	15	1210	P. tiles
Sanitryware waste	Kaolin	15	1145	W. tiles
Ceramic Sludge	Proportionally replaced raw materials	10	1160	W. tiles
Ceramic Sludge	Proportionally replaced raw materials	20	1180	F. tiles
Sewage Sludge	Proportionally replaced raw materials	60	1210	S. tiles
Sewage Sludge	---	70	980	G. tiles

(*Continued*)

Table 10.1 Name of waste used in the composition of tiles [10]. (*Continued*)

Name of wastes	Replacement of minerals	Atomic weight (%) of waste	Firing temperature (°C)	Types of tiles
Sewage Sludge	Proportionally replaced raw materials	7	1150	F. tiles
EAF Slag	---	40	1150	F. tiles
Blast furnace slag	Kaolin, limestone	33	1136	W. tiles
Basalt slag	Feldspar	5	1150	P. tiles
Hard dust rock	Feldspar	40	900	R. tiles
Red mud	---	66, 8	1180	F. tiles
Glass waste	Proportionally replaced raw materials	41	1080	St. tiles
Iron ore tailings	Feldspar	66	1200	P. tiles
Coffee husk ask	Feldspar	10	1180	F. tiles
Fish bone ash	Feldspar	10	1175	St. tiles
Rice husk ash	Clay	10	850	R. tiles

*Note:- R. tiles means roof tiles, F. tiles means floor tiles, St. tiles means stoneware tiles, P. tiles means porcelain tiles, W. tiles means wall tiles, G. tiles means glaze tiles, S. tiles means split tiles.

these materials, they form tiles. Sintering temperature ranges between 1100-1150 °C [13]. Teo et al. (2014) [14] have also prepared ceramic floor tiles using 40 wt.% of EAF slag with 20 wt.% silica, 10 wt.% feldspar, and 30 wt.% ball clay, and sintered at 1150°C. Ozturk and Gultekin (2015) [15] have applied BFS for preparing the wall tiles and concluded that incorporation up to 33 wt.% of BFS in the wall composition results in an improvement of strength by ~ 25%.

Granite dust waste is produced during rock mining, blasting and crushing of rocks. Granite contains high amount of SiO_2 and Al_2O_3 mixed with some amount of fluxes (Na_2O & K_2O) and coloring compound (Fe_2O_3) [16]. So far, this waste can also be considered as an alternative to conventional raw materials required for producing ceramics. Granite waste is found suitable due to low plasticity which reduces the possibility of dimensional defects. It is an alternate of feldspathic ingredients which make glassy phases at lower temperatures for fabrication of floor tiles [17]. Pazniak et al. (2018) [16] have studied effect of granitic rock and basalt waste incorporation into porcelain tiles. Addition of granitic and basalt rock wastes in place of feldspar exhibits the possibility of their limit in tiles industry as fluxes. Sample with 5 wt.% of basalt and sintered at 1150°C, show the same properties like the industrial porcelain tiles

Nowadays, waste glass powder is another material used to replace conventional fluxing materials in tiles manufacturing. Addition of glass causes decrement of sintering temperature [18, 19]. Gualtieri et al. (2018) [20] have developed low-temperature stoneware tiles by incorporation of waste glass in base triaxial composition. By choosing compositions promote modification of sintering temperature. Tile compositions contain 41 wt.% of glass waste (34 wt.% borosilicate and 7 wt.% soda-lime-silica glass) and 59 wt.% of traditional raw materials (i.e. 16 wt.% quartz, 28 wt.% clay, 15 wt.% feldspars) has reduced sintering temperature by ~135°C (1080°C instead of 1215°C).

10.3 Glass

Glass is a non-metallic, non-crystalline or amorphous inorganic solid. It has many engineering applications for our day to day life. There are many applications such as window panels, bottles, optoelectronics, transports, fiber optic cables, table-wares etc., are prepared from glass. Glasses are mostly made with sand (for SiO_2), feldspar (for SiO_2, Al_2O_3, Na_2O or K_2O), soda ash (for Na_2O), limestone (for CaO) and cullet (recycled glass) [21–29].

Objective of glass industries is to use glass which is economical and available locally. Glass manufacturers are producing glass which has low-cost.

10.4 Glass-Ceramic (GC)

Glass-ceramics is a multi-component material. GC has excellent properties in comparison to ordinary [30]. Some silica-based solid wastes i.e., fly ash, steel plant slag, filter dust, different types of sludge, and mud from metal processing industries. Glass cullet is used as raw material for the production of glass ceramics [31]. Both blast-furnace slag (BFS) and fly ash/pond ash are found more promising wastes for preparation of GC. Glass ceramics are used as building materials for aesthetics and construction components.

10.5 Mullite

Mullite is one of the oxide ceramics. Its molecular formula is $3Al_2O_3.2SiO_2$ and composition is ~28 wt% SiO_2 and 72 wt% Al_2O_3 [32, 33]. Mullite shows high refractoriness (>1700°C), low density (~3.17 g/cm³), high creep resistance, chemical resistance, good thermal stability, high modulus of rupture (HMoR), low dielectric constant, good corrosion resistance and low coefficient of thermal expansion [32–36]. Because of these properties, mullite is chosen for engineering applications [37]. This mineral is found in places i.e., Mull Island in Scotland. According to this Island, the name of mineral is called Mullite [32, 37]. Mullite can also be prepared by sintering of minerals, which contains requisite amount of silica and alumina. Usually, sintering temperature is at above 1300°C [32–34].

Presently, mullite is also prepared from low cost waste materials. These wastes are usually fly ash, pond ash, slag, etc.

10.6 Wollastonite

Wollastonite is a calcium silicate ($CaSiO_3$) mineral. It has some engineering properties i.e., low thermal conductivity, low dielectric value, low dielectric loss, good chemical stability, good corrosion resistance, low thermal expansion, fluxing properties, and whiteness [38]. Because of these properties, wollastonite is used by ceramic industries, metallurgical industries, paints industries, plastics industries, constructions industries, chemicals industries, etc [39, 40]. Nearly 30-40% wollastonite is consumed by ceramic industries in worldwide. $CaSiO_3$ improves ceramics performances.

In the presence scenario, this mineral have demand world-wide. However, wollastonite is not found in many countries. Present survey has not shown availability of enough $CaSiO_3$.

Hence, it is desirable to prepare wollastonite from low-cost waste ingredients via different synthetic route.

10.7 Cordierite

Cordierite ($Mg_2Al_4Si_5O_{18}$) is a well-known oxide ceramics. It has novel properties i.e., low value of thermal coefficient, thermal stability at elevated temperature, superior insulating property, low value of dielectric constant, good chemical resistance behavior, and high durability. It is a promising material for fabricating components in different engineering applications (like thermal insulators, furnace refractories, membranes, filters, integrated circuit boards, and catalysts) [41–43]. Pure form of cordierite synthesis is possible by sophisticated instrument. However, it is not practicable because of higher cost. Cost of raw materials of Cordierite is high.

Now-a-days, scientific communities have focused their research on preparation of low-cost cordierite from waste materials.

10.8 Silicon Carbide

Silicon carbide (SiC) is one of the encouraging non-oxide ceramics. It has excellent corrosion/wear resistance, good thermal stability, high mechanical strength, high hardness value, chemical inertness, wide band gap, and unique optical property. Because of the above properties, SiC is very useful in different engineering applications i.e., grinding media, heating elements, electronic devices, optic devices, catalyst support materials, reinforcement in ceramics, polymer and metal-matrix composites [44–46].

In last three decades, rice husk (RH) has acquired importance as a starting ingredient for the synthesis of SiC (particles and whiskers). Thus, Cutler *et al.* (1974) [47] have first introduced RH for the fabrication of SiC.

10.9 Silicon Nitride

Silicon nitride (Si_3N_4) is a high-temperature application non-oxide ceramic material. It has low thermal expansion coefficient, higher thermal shock and high creep resistance (than other ceramics), good corrosion resistance, and high temperature strength. This is because it is potential with favorable candidate for high-temperature structural ceramics applications [48, 49]. Therefore, many processes have been adopted to prepare Si_3N_4 powder and are not economical viable. Carbothermal-nitridation of silica route is economical and Si_3N_4 powder is prepared [50–53].

10.10 Ceramic Membranes

Membrane is a favorable candidature in separation technology. It is widely used in different technological applications (i.e., drink water production, wastewater treatment, gas purification and alkaline or acidic media separation) [54, 55]. In few decades, ceramic based membranes are developed and are more focused (i.e., if compared to other materials based membrane). It has many more advantages and are good chemical resistance, thermo-mechanical stability, high separation efficiency, easy clean regeneration, anti-fouling performance and long lifetime [56, 57]. Hence, Scientists or research community are trying to develop low cost ceramic membranes by utilizing wastes (or low cost ingredients).

10.11 Challenges

The world is experiencing environmental pollution due to generation of industrial wastes, by-products, gases, etc. Waste produced from thermal power plant and other sources cause global warming. Therefore, scientists/Engineers/Researchers are facing enough challenges to solve these problems. One way to tackle pollution is to control industrial pollution. Other way is power generation by renewable sources (such as solar/nuclear/wind/etc). This will make carbon free atmosphere. Different ways are thought by materials scientists. They are looking for inventing new materials at lower cost. In other words, carbon free world will create a situation to be congeners to human life.

Cement is a very important constructional material in the 20th century. This material can be produced in high energy process. It will cause an increase global temperature and increase in global pollution. If cement can be replaced by other materials, then pollution can be controlled to a great extent.

Present work has attempted to make a new materials i.e., high resistance ceramic for different applications in ceramic sectors. As started earlier, many scientists are working on ceramic production using Fly ash and other waste resource materials.

10.12 Opportunity

Replacement of porcelain based ceramic by industrial waste will be a boon to present scenario. The other important aspect is economic consideration.

Encouragement of entrepreneur activities will help to solve employment problems.

In addition, industrial waste based ceramic will immensely benefit the modern world.

While thinking of such materials one must consider ease of production and availability of cheaper raw materials.

All these points will provide opportunity to the entrepreneur for starting business. This will help the country to develop at a faster pace.

In conclusions, following points are drawn from the above;

i. Industrial waste based ceramic is a cost effective materials since raw materials are available as waste. Therefore less cost is involved.
ii. Utilization of waste materials will protect land.
iii. Simplicity in production process, where high temperature is needed.
iv. Ease of production.
v. Replacement of costlier conventional ceramics.
vi. Solve employment problem by increase of entrepreneur activities.

This will help to develop a nation.

10.13 Conclusions

People pay attention to environmental protection in the 21st century. Higher demands are being raised for environmental protection by planning strategy of sustainable development. Reducing environment pollution and recycling of ceramic wastes are helpful for environmental protection. Possibility for use of ceramic waste in concrete can be seen from the precedent usage of fly ash which is a waste.

Our country is the world's largest ceramics producer. Countries need large tonnage of concrete. It is needed to accelerate our development at a higher speed. Looking to the availability of raw materials, non-renewable resources are close to the red line. If waste can be used fully, environment damage will be reduced. Again avoiding conventional method of ceramic production and usage of waste materials as raw materials will enable benefit to countries economy. By this process, we can avoid environmental crisis. Hence, sustainable development economy of society can be reached.

Different possible use of waste are given below;

The waste derived nanomaterials (i.e., Fe_3O_4, MFe_2O_4 (M- Mn, Cu, Zn, Ag), magnetic biochar, $MgCr_2O_4$, Cr_2O_3, CuO, Copper ferrite, etc.) synthesized by physico-chemical methods have wide application in environmental remediation.

Waste may be utilized in the building ceramics production as Catalyst.

The production of stable inorganic pigments only from wastes is feasible.

Binary and ternary combinations are validated for making glass-ceramic materials possessing valuable properties. These glass-ceramics may be used for different purposes such as floor and wall tiles, bench tops, sewer pipes and many others.

The review should supply a comprehensive source of info to those involved in this area, both of academia and industry for finding new ways to recycling wastes in production of ceramics with viable and alternative technique. However, more investigation is required in regarding to sustainable development in ceramics with technology transfer from academica to industries.

Acknowledgments

First author would like to thank Prof. Munesh Chandra Adhikary, PG Council Chairman, Fokir Mohan University, for his invaluable guidance, advice, and constant inspiration throughout the entire program. First author thanks Mr. Mukteswar Mohapatra in Fokir Mohan University for his support. The authors convey their sincere thanks to GIET, University Gunupur, Rayagada, Odisha, India for providing Lab facilities to do the research work. The authors also thank the CRF, IIT Kharagpur for providing their testing facilities.

References

1. J. Temuujin, A. van Riessen, K. J. D. MacKenzie, (2010), Preparation and Characterization of Fly Ash based Geopolymer Mortars, *Construction and Building Materials*, Vol. 24, October, pp. 1906-1910, https://doi.org/10.1016/j.conbuildmat.2010.04.012.
2. N. B. Singh, (2018), Fly Ash-Based Geopolymer Binder: A Future Construction Material, *Minerals*, Vol. 8, 299, https://doi.org/10.3390/min8070299.
3. Nabila Shehataa, O. A.Mohamed, Enas Taha Sayed, Mohammad Ali Abdelkareem, A. G. Olabi, (2022), Geopolymer Concrete as Green Building Materials: Recent Applications, Sustainable Development and Circular Economy Potentials, *Science of The Total Environment*, Vol. 836, https://doi.org/10.1016/j.scitotenv.2022.155577.
4. Y. Sun, (2006), The Present Study State and Problems to be Solved on Recycled Concrete in China, *Concrete*, Vol. 4, pp. 25-28.
5. H. Mizuguchi, (1998), A Review of Environmentally Friendly Concrete, *Journal of Concrete*, Vol. 36.
6. X. Wang, (2012), Experimental Study of the Preparation of Ecological Concrete using Waste Fly Ash, Master's thesis, School of Civil Engineering of Shandong Jianzhu University.
7. (a). A. M. Hassan, H. Moselhy, and M. F. Abadir, (2019), The Use Of Bagasse In The Preparation of Fireclay Insulating Bricks, *International Journal of Applied Ceramic Technology*, Vol.16, pp. 418-425. (b). M. Sutcu, S. Akkurt, and A. Bayram, (2012), Production of Anorthite Refractory Insulating Firebrick from Mixtures of Clay and Recycled Paper Waste with Sawdust Addition, *Ceramic International*, Vol. 38, pp. 1033-1041. (c). A. K. Mandal, H. R. Verma, O. P. Sinha, (2017), Utilization of Aluminum Plant's Waste for Production of Insulation Bricks, *Journal of Clean Production*, Vol. 162, pp.949-957. (d). F. Fan, Z. Liu, G. Xu, (2018), Mechanical and Thermal Properties of Fly Ash based Geopolymers, *Construction Building Materials*, Vol. 160, pp. 66-81. (e). R. Sukkae, S. Suebthawilkul, B. Cherdhirunkorn, (2018), Utilization of Coal Fly Ash as a Raw Material for Refractory Production, *Journal of Metals, Materials and Minerals*, Vol. 28, pp.116-123.
8. N. M. Khalil, Y. Algamal, and Q. M. Saleem, (2018), Exploitation of Petroleum Waste Sludge with Local Bauxite Raw Material for Producing High-Quality Refractory Ceramics, *Ceramic International*, Vol. 44, pp. 18516-18527, https://doi.org/10.1016/j.ceramint.2018.07.072.
9. Dublin. Global ceramic tiles markets, (2019), Available at. https://www.globenewswire.com/news-release/2019/02/19/1736582/0/en/Global-Ceramic-TilesMarkets-2011-2018-2019-2024.html Accessed on 2019 Jun 30.
10. Sk S. Hossain and P.K. Roy, (2020), Sustainable Ceramics Derived from Solid Wastes: A Review, *Journal of Asian Ceramic Societies*, Vol. 8, pp. 984-1009, https://doi.org/10.1080/21870764.2020.1815348.

11. N. Chandra, P. Sharma, and G. L. Pashkov, (2008), Coal Fly Ash Utilization: Low temperature Sintering of Wall Tiles, *Waste Manage*, Vol. 28, pp. 1993-2002.
12. H. X. Lu, M. He, and Y. Y. Liu, (2011), A Preparation and Performance Study of Glass-Ceramic Glazes Derived from Blast Furnace Slag and Fly Ash, *Journal of Ceramic Processing Research*, Vol.12, pp. 588-591.
13. R. Sarkar, N. Singh, and S. K. Das, (2010), Utilization of Steel Melting Electric Arc Furnace Slag for Development of Vitreous Ceramic Tiles, *Bulletin of Materials Science*, Vol. 33(3), pp. 293-298.
14. P. T. Teo, A. A. Seman, and P. Basu, (2014), Recycling of Malaysia's Electric Arc Furnace (EAF) Slag Waste into Heavy-Duty Green Ceramic Tile, *Waste Management*, Vol. 34, pp. 2697-2708.
15. Z. B. Ozturk and E. Gultekin, (2015), Preparation of Ceramic Wall Tiling Derived from Blast Furnace Slag, *Ceramic International*, Vol. 41, pp. 12020-12026.
16. A. Pazniak, S. Barantseva, and O. Kuzmenkova, (2018), Effect of Granitic Rock Wastes And Basalt on Microstructure and Properties of Porcelain Stoneware, *Materials Letter*, Vol. 225, pp. 122-125.
17. P. Torres, R. S. Manjate, and S. Cuaresma, (2007), Development of Ceramic Floor Tile Compositions based on Quartzite and Granite Sludges, *Journal of the European Ceramic Society*, Vol. 27, pp. 4649-4655.
18. A. P. Luz and S. Ribeiro, (2007), Use of Glass Waste as a Raw Material in Porcelain Stoneware Tile Mixtures, *Ceramic International*, Vol. 33, pp. 761-765.
19. K. Kim, K. Kim, and J. Hwang, (2016), Characterization of Ceramic Tiles Containing LCD Waste Glass, *Ceramic International*, Vol. 42, pp. 7626-7631.
20. M. L. Gualtieri, C. Mugoni, and S. Guandalini, (2018), Glass Recycling in the Production of Low-Temperature Stoneware Tiles, *Journal of Cleaner Production*, Vol. 197, pp. 1531-1539.
21. J. Kaewkhao and P. Limsuwan, (2012), Utilization of Rice Husk Fly Ash in the Color Glass Production, *Process Engineering*, Vol. 32, pp. 670-675.
22. M. Erol, S. Kucukbayrak, A. Ersoy-Mericboyu, (2007), Characterization of Coal Fly Ash for Possible Utilization in Glass Production, *Fuel*, Vol. 86, pp. 706-714.
23. J. Sheng, B. X. Huang, J. Zhang, (2003), Production of Glass from Coal Fly Ash. *Fuel*, 82(2):181-185.
24. H. S. Park and J. H. Park, (2017), Vitrification of Red Mud with Mine Wastes through Melting and Granulation Process-Preparation of Glass Ball, *Journal of Non-Crystalline Solids*, Vol. 475, pp. 129-135.
25. I. Kashif and A. Ratep, (2018), Preparation and Characterization of Oxide Glass from Sugar Cane Waste, *Silicon*, Vol. 10, pp. 2677-2683.
26. Y. Xu, Y. Zhang, and L. Hou, (2014), Preparation of $CaO-Al_2O_3-SiO_2$ System Glass from Molten Blast Furnace Slag, *International Journal of Minerals, Metallurgy and Materials*, Vol. 21, pp.169-174.

27. E. C. Hammel, L. R. Ighodaro, and O. I. Okoli, (2014), Processing and Properties of Advanced Porous Ceramics: An Application based Review, *Ceramic International*, Vol. 40, pp. 15351-15370.
28. R. Ji, S. Wu, and C. Yan, (2017), Preparation and Characterization of the One-Piece Wall Ceramic Board by using Solid Wastes, *Ceramic International*, Vol. 43, pp. 8564-8571.
29. X. Fang, Q. Li, T. Yang, (2017), Preparation and Characterization of Glass Foams for Artificial Floating Island from Waste Glass and Li_2CO_3, *Construction and Building Materials*, Vol. 134, pp. 358-363.
30. W. Holand and G. H. Beall, (2013), Handbook of Advanced Ceramics, Chapter 5.1 - Glass-Ceramics. Elsevier- Vol. 371. New York, United States.
31. R. D. Rawlings and J. P. Wu, (2006), Boccaccini AR. Glass-Ceramics: Their Production from Wastes-A Review, *Journal of Materials Science*, Vol. 41, pp. 733-761.
32. H. Schneider, K. Okada, J. A. Pask, (1994), Mullite and Mullite Ceramics, Chichester, UK: Wiley.
33. H. Schneider, J. Schreuer, and B. Hildmann, (2008), Structure and Properties of Mullite-A Review, *Journal of the European Ceramic Society*, Vol. 28, pp. 329-344.
34. Y. F. Chen, M. C. Wang, M. H. Hon, (2004), Phase Transformation and Growth of Mullite in Kaolin Ceramics, *Journal of the European Ceramic Society*, Vol. 24, pp. 2389-2397.
35. V. Mandić, E. Tkalčec, and J. Popović, (2016), Crystallization Path Way of Sol-Gel Derived Zinc-Doped Mullite Precursors, *Journal of the American Chemical Society*, Vol. 36, pp. 1285-1292.
36. R. Zhang, X. Hou, and C. Ye, (2017), Enhanced Mechanical And Thermal Properties of Anisotropic Fibrous Porous Mullite Zirconia Composites Produced using Sol-Gel Impregnation, *Journal of Alloys and Compounds*, Vol. 699, pp. 511-516.
37. I. A. Aksay, D. M. Dabbs, and M. Sarikaya, (1991), Mullite for Structural, Electronic and Optical Applications, *Journal of the American Ceramic Society*, Vol. 74, pp. 2343-2354.
38. S. Sen, (1992), Ceramic White Wares: Their Technologies and Applications, 1st Ed. Oxford & IBH publishing CO. PVT. Ltd;. pp. 1-169, New Delhi, India.
39. X. Y. Liu, C. X. Ding, and P. K. Chu (2004), Mechanism of Apatite Formation on Wollastonite Coatings in Simulated Body Fluids, *Biomaterials*, Vol. 25, pp. 2007-2012.
40. R. L. Virta, (2009), Wollastonite, U.S. geological survey. Minerals Yearbook. https://s3-us-west-2.amazonaws.com/prd-wret/assets/palladium/production/mineralpubs/wollastonite/ myb1-2009-wolla.pdf, Accessed on 2019 Jul 11.
41. R. Bejjaoui, A. Benhammou, and L. Nibou, (2010), Synthesis and Characterization of Cordierite Ceramic from Moroccan Stevensite and Andalusite, *Applied Clay Science*, Vol. 49, pp. 336-340.

42. W. Wang, Z. Shin, and X. Wang, (2016), The Phase Transformation and Thermal Expansion Properties of Cordierite Ceramics Prepared using Drift Sands to Replace Pure Quartz, *Ceramic International*, Vol. 429, pp. 4477-4485.
43. D. Kuscer, I. Bantan, and M. Hrovat, (2017), The Microstructure, Coefficient of Thermal Expansion and Flexural Strength of Cordierite Ceramics Prepared from Alumina with Different Particle Sizes, *Journal of the European Ceramic Society*, Vol. 37, pp. 739-746.
44. A. Davidson and D. Regener, (2000), A Comparison of Aluminium-based Metal-Matrix Composites Reinforced with Coated and Uncoated Particulate Silicon Carbide, *Composites Science and Technology*, Vol. 60, pp. 865-869.
45. W. Yang, H. Araki, and C. Tang, (2005), Single-Crystal SiC Nanowires with a Thin Carbon Coating for Stronger and Tougher Ceramic Composites, *Advanced Materials*, Vol. 17, pp. 1519–1523.
46. G. Yang, H. Cui, and Y. Sun, (2009), Simple Catalyst-Free Method to the Synthesis of-SiC Nanowires and Their Field Emission Properties, *Journal of Physical Chemistry C*, Vol. 113, pp. 15969-15973.
47. I. B. Cutler, J. G. Lee, and N. Shaikh, (1974), UTEC A, Vol. 73, 157(A).
48. Herrmann M, Klemm H, Schubert C. (2000), Silicon Nitride based Hard Materials, *Handbook of Ceramic Hard Materials*, pp. 749–801, https://doi.org/10.1002/9783527618217.ch21.
49. M Wang, M Xie, and L. Ferraioli, (2008), Light Emission Properties and Mechanism of Low-Temperature Prepared Amorphous Sin Films. I. Room-Temperature Band Tail States Photoluminescence. *Journal of Applied Physics*, Vol. 104, 083504.
50. F. L. Riley, (2000), Silicon Nitride and Related Materials, *Journal of the American Ceramic Society*, Vol. 83, pp. 245-265.
51. T. Licko, V. Figusch, and J. Puchyova, (1992), Synthesis of Silicon Nitride by Carbothermal Reduction and Nitriding of Silica: Control of Kinetics and Morphology, *Journal of the European Ceramic Society*, Vol. 9, pp. 219-230.
52. L. Sun and K. Gong, (2001), Silicon-based Materials from Rice Husks and Their Applications, *Industrial & Engineering Chemistry Research*, Vol. 40 pp. 5861-5877.
53. C. Real, M. D. Alcala, and J. M. Criado, (2004), Synthesis of Silicon Nitride from Carbothermal Reduction of Rice Husks by the Constant-Rate-Thermal-Analysis (CRTA) Method, *Journal of the American Chemical Society*, Vol. 87, pp. 75-78.
54. M. Padaki, R. S. Murali, and M. S. Abdullah, (2015), Membrane Technology Enhancement in Oil-Water Separation, *A Review Desalination*, Vol. 357, pp. 197-207.
55. A. T. Mohammadi, (2007), A Review of the Applications of Membrane Separation Technology in Natural Gas Treatment, *Separation Science and Technology*, Vol. 34, pp. 2095-2111, https://doi.org/10.1081/SS-100100758.

56. H. Ramlow, R. K. M. Ferreira, C. Marangoni, (2019), Ceramic Membranes Applied to Membrane Distillation: A Comprehensive Review, International *Journal of Applied Ceramic Technology*, Vol. 16, pp. 2161-2172.
57. K. S. Ashaghi, M. Ebrahimi, and P. Czermak, (2007), Ceramic Ultra and Nanofiltration Membranes for Oilfield Produced Water Treatment: A Mini Review, *The Open Environmental Research Journal*, Vol. 1, pp. 1-8.

Index

Accelerated durability testing, 55–56
Accelerated mortar bar test, 53
Agate mortar, 206
Air-cooled BF slag (ACBFS), 42
Aircraft, geopolymer-based materials in, 231, 232
Air pollution, 29–30, 32, 170, 231
Albite, 218
Alccofine powder, 93
Alkali-activated-based geopolymer, 153
Alkali-activated FA/BFS, 153
Alkali-activated materials (AAMs), 199
Alkali activation of ceramic waste (AACW), 50, 76
Alkali dosage effects, 155
Alkaline sodium silicate, 48
Alkali silicon-oxoaluminate, 45
Alumina, 151, 153
Aluminium salt solution, 171–172
Alumino-silicate geopolymer, 45, 47, 49, 75, 201
Aluminosilicate hydrogels, 49, 75
Aluminum and silicon-bearing materials, GP preparation, 154
Aluminum powder, 174, 233
Aluminum silicates, activation of, 196
Amazonian kaolin, 171
Anji Bridge (China), 8
Anthracite, 26
Archaeological structures, with geopolymeric materials, 231, 232
Argon gas, 102, 103

Argon oxygen decarburization (AOD) slag, 48, 73
Asbestos in roofing components, 171
Ash/ashes, types, 18–33
  cigar ash, 19
  coal ash and fly ash, 21–24
  coconut shell ash, 21
  fly ash generation, 24, 25
  pond ash, 29–31
  pond ash management, importance of, 32–33
  pulverized fuel ash, various uses of, 31–32
  quarry dust, 20
  RHA, 19
  thermal power plant ashes, nature and composition, 24, 26–29
  volcanic ash, 19–20
  wood ash, 19
Atomic absorption spectroscopy (AAS), 213, 216

Backfilling material (BM), microstructure of, 154
Bagasse cellulose fiber, 173
Baghdad power plant, 52, 77
BALCO, 170
Bamboo fibers, 171
Bands (peaks), 219–220
Basalt fibers, 174–175
Basalt strands, 172
Basic oxygen furnace (BOF) slag, 40, 42
Beauvais cathedral, 11

BET (Brunauer, Emmett and Teller) surface areas of pond ashes, 93
BFS. *see* Blast furnace slag (BFS)
Binders, geopolymer, 228–229
Biochar, 48, 74
Bituminous coal, 26
Blaine air-permeability method, 35
Blast furnace slag (BFS), 40, 41, 42, 44
  alkali-activated, binding mechanism of, 153
  compressive strength tests, 153
  flexural strength test, 153
  geopolymer production from, 73, 74, 75, 76, 77
  GGBFS, 42, 49, 50, 74, 75, 76
Blowing agent, 233
BOF (basic oxygen steelmaking) slag, 42
Bottom ash, 23, 24, 29, 30, 92, 124–125
  chemical composition of, 125
  collection, 92
  engineering properties, 125
Bragg's law, 101
Brick(s), 12, 13, 125, 229, 232
  brick-making technology, 9
  burnt clay, 29
  eco-friendly, 48, 74, 174
  fired, 6, 8, 23
  with lime mortar, 9
  manufacture of, 31
  mold-made mud, 2
  mud-brick, 2, 4, 5
  production, 16
  weight loss of, 174
Bridges, organic matrix-based continuous fiber composites, 232, 233
Bronze Age, 3
Brown coal fly ash, 94
Bulk density, PA-based GP mortar/concrete, 207–211, 212, 213
Burnt clay bricks, production of, 29

Calcined kaolin, 48
Calcium aluminate cements, 49, 74
Calcium aluminum silicate hydrate (CASH), 153
Calcium electric arc ferronickel slags, 73, 76
Calcium hydroxide, 202
Calcium silicate hydrate (CSH), 153, 156
Carbon dioxide (CO_2), emission of, 73, 152, 196
Carbon fiber, 174–175
  -reinforced geopolymer composites, 233
Carbonic acid, 198
Carbon negative cement, 52, 78
Carl Zeiss Supra-40 Scanning Electron Microscope, 206
Casting,
  PA-based geopolymer products, 203, 204, 205
  PA/HCFC slag-based GP preparation, 157
  PA/jute fiber-based GP, 179, 180, 181
  preparation of geopolymer from PA, 99, 100
Cast iron, use of, 16
Caustic soda. *see* Sodium hydroxide
Cellulose, 171
Cement(s); *see also* High carbon ferrochrome (HCFC) slag-based GP cementitious materials; Pond ash (PA)-based GP cementitious materials; Pond ash (PA)–jute fiber-based GP cementitious materials
  alkali-activated, 47, 72, 228
  -based materials, corrosion of, 196, 198, 199
  biochar with, 48, 74
  blended, 124
  calcium aluminate, 49, 74
  carbon negative, 52, 78

cementing agent, 26, 27
FABC, 49, 75
factories, 9
flowable slurry along with, 32
formulation of, 49
hydrates and aggregate, 197
in acid media, 201
in concrete, 73
industry, slag in, 40, 41
inorganic polymer, 72
magnesium chloride phase of, 77
magnesium oxy chloride (sorel)-based, 52, 77
magnesium oxy sulfate-based, 52, 77
mortar, durability behavior of, 55
PC. *see* Portland cements (PC)
pore radii in, 197
production, 46, 72, 92, 196, 234
replacement of, 29, 47, 48, 74, 124, 125, 148, 154, 161, 200, 202, 207, 228, 229, 234
resource material for preparing, 31
RHA with, 19
SCMs and, 200
seawater resistant, 55
self-cementing behavior, 27
slag, 42, 44
sodium sulfate-activated slag, 51, 77
SSC, 49, 74
for storage, 229
thermal energy, 53
using ferrochrome slag (FS), 155
Ceramics, geopolymer, 230, 231
Characterization of geopolymer (GP), 155, 159, 171, 172, 197
DSC. *see* Differential scanning calorimetry (DSC)
FTIR. *see* Fourier-transform infrared spectroscopy (FTIR)
IR, 100
PA-based GP mortar/concrete (before and after) corrosion, 206–207

PA/HCFC-based geopolymeric material, 158–159
pond ash, 32
of prepared samples, 127
SEM. *see* Scanning electron microscopy (SEM)
techniques, 48, 49, 74
TGA. *see* Thermogravimetric analysis (TGA)
XRD. *see* X-ray diffraction (XRD)
Chartres Cathedral, 11
China,
ancient, 8–9
coal deposits in, 21, 23
Chloride-induced corrosion durability of GP, 201
Chopped fibers, 179, 181, 186
Cigar ash, 19
Cinder. *see* Slag(s)
Class C fly ash, 26, 27–29
Class F fly ash, 26–27, 28, 155
Clinker lumps, 29
Coade stone, 16
Coal,
combustion, 23, 24, 25
consumption, 21, 23
fly ash slurry spill, 30
fossil fuel, 21
Coal ash(es), 21–24, 29–31
chemical composition, 36
chemical properties of, 35–36
compaction behavior, 36–37
leaching behavior, 37–38
mineralogical phases, 36
morphology of, 33
permeability, 37
physical characteristics of, 33–38
solubility of solids, 36
specific gravity of, 33
specific surfaces of, 35
strength behavior, 37
surface area of, 33
utilization, 38–39

Coal fly ash, for preparation of GP, 154
Coconut shell ash, 21
Coir fibers, residual, 171
Compacted mass, formation, 141
Compaction, 33–34
    behavior, of coal ash, 36–37
    PA-based geopolymer products, 203, 204, 205
    PA/HCFC slag-based GP preparation, 157
    PA/jute fiber-based GP, 179, 180, 181
    preparation, PA-based GPs, 99, 100
Compression testing,
    of as-prepared geopolymer samples, 127
    PA-jute fiber-based GP, 172
Compressive strength,
    of corroded geopolymer composite, 200–201
    PA-based GP mortar/concrete, 206, 212, 214, 215
    of PA/HCFC slag GP materials, 153, 154–154, 156, 158, 159, 160
    PA-jute fiber-based GP, 173, 174, 182–185, 186
Concrete,
    cracking in, 54
    deterioration process of, 53–54
    dissolution mechanism of steel in, 54
    durability of, 53–55
    electrochemical process, 54
    mechanism of steel corrosion, 54
Construction materials, historical development of,
    accelerated durability testing, 55–56
    ash/ashes, types, 18–33
        cigar ash, 19
        coal ash and fly ash, 21–24
        coconut shell ash, 21
        fly ash generation, 24, 25
        pond ash, 29–31
        pond ash management, importance of, 32–33
        pulverized fuel ash, various uses of, 31–32
        quarry dust, 20
        RHA, 19
        thermal power plant ashes, nature and composition, 24, 26–29
        volcanic ash, 19–20
        wood ash, 19
    chronological development, 2–18
        ancient China, 8–9
        ancient Egypt, 4–5
        ancient Greece and Rome, 5–8
        ancient Mesopotamia, 4
        Copper Age and Bronze Age, 3
        eighteenth century, 15–16
        Iron Age, 3–4
        Middle Ages, 9–11
        Neolithic Age, 2–3
        nineteenth century, 16–17
        Renaissance, 11–15
        seventeenth century, 15
        Steel Age, 3–4
        twentieth century, 17–18
    coal ash(es)
        physical characteristics of, 33–38
        utilization, 38–39
    durability of concrete, 53–55
        deterioration process, 53–54
        dissolution mechanism of steel in, 54
        electrochemical process, 54
        mechanism of steel corrosion, 54
    geopolymers, 45–53
        activated, preparation of, 48
        alkali activation mechanism for, 45
        applications of, 47
        at ambient temperature, 49–50
        by alkaline activation, 49
        chemical composition of, 45
        constituents of, 46–52
        defined, 45

fabricated plant fiber-based, 48
fire-resistant, 51
GPC, 47
hybrid inorganic-organic, 50
in alkaline solution, 50
kaolin, 51
metakaolin-based, 51
non-colored and colored, 50
properties, 45, 52–53
reaction, schematic
   representation of, 46
sol-gel method for, 48
through alkali-silicate activation, 50
waste glass-based, 49
slag, 39–45
  ACBFS, 42
  BFS, 41, 42
  BOF, 42
  chemistry, 39
  converter, 44
  electric arc furnace slag, 44
  ferrous and non-ferrous smelting processes, 39
  GBFS, 42
  generation of, 40–44
  GGBFS, 42
  iron and steel slag, 43, 44
  LD or steel slag, 42
  optical photo images, 41
  properties and utilization, 44–45
  reducing, 44
  rock, 42
  synthetic, 39
Converter slag, 44
Copper Age, 3
Copper mine tailings, 48, 74
Corchorus capsularis, 178
Corchorus olitorius, 178
Corrosion,
  behavior in different corrosive medium, 53
  chloride-induced corrosion durability of GP, 201
  electrochemical process of, 199
  HCl corrosive media, 53
  inhibitors, 198
  in M30 grade geopolymer concrete, 201
  in sulphate and acid media, 200
  mechanism of, 199
  mechanism of steel, 54, 55
  of OPC concrete, 198
  of pond ash (PA)-based geopolymer products
    causes of, 197
    characterizations, 206–207
    chemicals and materials, 203
    comparison of weight change and bulk density, 207–211, 212, 213
    compositions of PA-based concrete mixtures, 205
    compositions of PA-based mortar mixtures, 204
    concrete preparation process, 204–205
    curing process, 204, 205
    experimental details, 203–220
    mechanical property of GP mortar/GP concrete, 211, 212, 214, 215
    molding, casting and compaction, 203, 204, 205
    mortar preparation, 203–204
    overview, 196–202
    preparation of, 203–205
    raw material preparation, 203, 205
    results and discussion, 207–220
  of steel reinforcement, 198
  potential measurement, 200
  property of steel-reinforced concrete structures, 55
  rate, 200
  reinforced bar coated with GP paste, 198
  resistance of high-strength reinforced concrete, 200

204 INDEX

types of, 198
Cotton, 171
Cracking in concrete structures, 54
Crack widths, 174
Crushed residual waste glass (RWG), 74
Curing,
  compacted mass formation, 141
  green materials, 202
  microwave-oven, 93
  of GP paste, 198
  PA, preparation of geopolymer from, 100
  PA-based geopolymer products, 204, 205, 207
  PA/HCFC slag-based GP preparation, 157
  PA/jute fiber-based GP, 174, 179, 180, 181–182
  temperatures, 105, 106, 128–132, 207
  thermal, 94
  times, 128–132, 156, 207
    length of, 106, 107, 108, 109
    variations of, 106, 107

Dan River coal ash spill, 30
Desiccator method, 35
Dicalcium silicate, 44
Differential scanning calorimetry (DSC),
  characteristic properties of FP pastes, 141, 146, 147
  HCFC slag-based GPs, 156, 158, 159
  PA-based GP mortar/concrete, 207
  PA-based GPs, 102, 103–104, 112, 115
  PA-jute fiber-based GP, 175, 182
Dog bone tensile testing, 174
Dolomite, 52
DSC. *see* Differential scanning calorimetry (DSC)

Dumped fly ash, 24
Durability,
  accelerated durability testing, 55–56
  of AAMs, 199
  of cement mortar, 55
  of concrete, 53–55
  chloride-induced corrosion durability of GP, 201
  test in sulfuric acid, 93

Eco-friendly GP composite, 48, 74, 171–172, 174
EDX. *see* Energy-dispersive X-ray (EDX) analysis
Egypt, ancient, 4–5
Eighteenth century, constructional materials, 15–16
Electric arc furnace (EAF) slag, 40, 44
Electron microscopy techniques, 101–102
Electron tomography, pores in geopolymer, 94
Embankments, design, 30
Energy-dispersive X-ray (EDX) analysis,
  as-received PA and prepared PA-based GP, 141, 146
  characterization of prepared samples, 108, 109, 110, 127
  HCFC slag-based GPs, 158, 159–161
  PA-based GP mortar/concrete, 206
  PA/jute fiber-based GP, 181
Environmental pollution, 46, 72, 125, 152, 234
Epoxy-based organic resins, 49
Ettringite, 51, 75, 77
Eucalyptus waste, 171

Fabricated plant fiber-based geopolymer, 48
Fall cone method, 35

Ferrochrome slag (FS). *see* High carbon ferrochrome (HCFC) slag-based GP
Field emission scanning electron microscope (FESEM),
 as-received pond ash sample, 141, 146
 characterization of prepared samples, 101–102, 127
 geopolymer cement using HCFA, 202
Fired clays, 6
Fire hazards, 233
Fire-resistant geopolymers, 51, 77
Fire-resistant matrix composite materials, 231, 232
Flame atomic absorption spectrometry (F-AAS), 206
Flammability, of GP materials, 231, 232, 233
Flexural strength,
 of PA/HCFC slag GP materials, 153, 154–155
 PA-jute fiber-based GP, 173, 174
Florence cathedral, dome of, 14, 15
Flushing gas stream, fluctuations in, 102
Fly ash (FA), 21–24
 alkali-activated, binding mechanism of, 153
 brown coal, 94
 chemical composition of, 26–29
 classes of, 26–29
 compressive strength tests, 153
 different mass ratios of, 172
 dry density of, 37
 flexural strength test, 153
 for plant growth, 34
 generation, 24, 25, 26, 92
 low-calcium, 47, 72
 PFA, 31–32
 pH value, 36
 pond ash *vs.*, 30–31
 pozzolanic behavior of, 31
 solubility of solids, 36
 strength of, 36
Fly ash-based geopolymer, 76
Fly ash-based geopolymer concretes (FAGPC), 154
Fly ash belite cement (FABC), 49, 75
Foaming agent, 233
Fossil fuels, 21, 22
Fourier-transform infrared spectroscopy (FTIR),
 geopolymer cement using HCFA, 202
 HCFC slag-based GP, 156, 159
 nature of the geopolymeric reaction using, 154
 PA-based GP mortar/concrete, 206, 218, 219–220
 PA-based GPs, 100, 113, 127
 PA-jute fiber-based GP, 175, 182, 186, 187–188
Fragmentation, process of ash formation, 20
Free swell index, 34–35
FTIR. *see* Fourier-transform infrared spectroscopy (FTIR)

Gas adsorption method, pores in geopolymer, 94
Geopolymer concrete (GPC), 47, 78
Geopolymer foam concretes (GFCs), 50, 76
Geopolymer protective coatings (GPCs), 229–230
Geopolymers (GPs), 45–53
 applications of, 47
 chemical composition of, 45
 constituents of, 46–52
 defined, 45
 formation mechanism, 78–81
 materials

applications, 228–233
benefits of, 234–235
challenges, 234
opportunity, 234–235
overview, 228–233
strength, 228
mechanical properties of, 128, 141–147, 171, 202
micrographs of, 108, 109
overview, 72–78
parameters of, 78
preparation, from pond ash, 98–100, 126
properties, 45, 52–53
synthesis, 73, 74, 75–78
Glass fiber, 174
Glass for architectural purposes, 7–8
Glass panes, 16
Global warming, 46, 47, 72, 73, 196, 234
GPs. *see* Geopolymers (GPs)
Grain size distribution, 34
Granulated blast furnace slag (GBS or GBFS), 42, 44
Great Pyramid of Giza, 4
Great Wall of China, 8–9
Greece, ancient, 5–8
Greenhouse gas effect, 46, 47, 72, 73, 196
Ground granulated blast furnace slag (GGBFS), 42, 49, 50, 155
corrosion rate for, 201
geopolymer production from, 74, 75, 76, 154
Groundwater pollution, 32
Gypsum, 37, 49

HCFC. *see* High carbon ferrochrome (HCFC) slag-based GP cementitious materials
Heating, weight loss of GP cured samples and, 146, 147
Heat insulation performance of GP coating, 230

Hibiscus cannabinus, 178
High-calcium fly ash (HCFA), 202
High carbon ferrochrome (HCFC) slag-based GP cementitious materials,
experimental details, 156–163
characterizations of PA/HCFC-based geopolymeric material, 158, 159
PA/HCFC-based geopolymeric mortar and concrete, 158, 159
PA/HCFC slag-based GP preparation, 157
results and discussion, 159–163
source of materials, 156–157
overview, 152–156
Hollow glass microspheres, 230
Hydraulic conductivity, in soil, 34
Hydrogen peroxide (H_2O_2),
as blowing agent, 233
solution, 76
Hydro sodalite, crystalline phase of, 218

Ilmenite smelting, 39
Imhotep, 5
Index properties, in geotechnical engineering practice, 35
India,
coal ashes, 36
coal deposits in, 21, 23
demand for energy in, 32
steel industry in, 40
thermal power plants in, 124
Industrial wastes,
bearing aluminosilicate minerals, 230
dumping and landfilling, 170
GPC from, 229
GP production from, 72, 74, 76, 77, 228
Inhibitors, corrosion, 198
Insulating materials, 233
Ionic tetrahedral coordination, 79

Iron, use, 15, 16
Iron Age, 3–4, 6
Iron aluminum oxides, amorphous, 36
Iron Bridge at Coalbrookdale, 16
Iron slag, 40, 43, 44

Jade gate pass (Yumenguan), 9
Jiayuguan's Great Wall, 9
Jute fiber(s), 175–176
 -based GP, PA and. *see* Pond ash
  (PA)–jute fiber-based GP
  chemical composition, 178
  physical properties of, 176–178
  uses of, 176, 177

Kaolin, 51, 76
Kaolinite, 76
King's College Chapel, 11
Kingston fossil plant, 30
Kizhi (Russia), church in, 12, 13
Korogho church in Georgia, 10, 11

Labor, in the Renaissance, 13
LD (Linz-Donawitz) slag, 42, 44, 45
Leaching behavior, 37–38
Lighthouse of Alexandria, 6
Lignite, 27
Lime reactivity, 36
Limestone fillers (F-Lime), 49, 74
Linz-Donawitz (LD) slag, 42, 44, 45
Loba chemicals, 126, 175
Louvre in Paris, roof of, 16
Lye. *see* Sodium hydroxide

Magma, 20
Magnesium oxy chloride (sorel)-based
 cements, 52, 77
Magnesium oxy sulfate-based cements,
 52, 77
Magnesium sulfate, 154
Manganese oxides, 36
Masonry techniques of ancient Greece
 and Rome, 5–6
Media, corrosive, 213, 214, 216, 218

Menkaure pyramid, 4
Merck (India), 126, 156–157, 175
Mercury porosimetry method, pores
 in geopolymer, 94
Merwinite, 49
Mesh sizes, variations of, 105
Mesopotamia, ancient, 4
Metakaolin, 47, 48, 49, 51, 72, 73, 76,
 173
 -based geopolymeric mortar, 77
 synthesis of geopolymer from,
  230
Metallurgical slag, 51, 77
M30 grade geopolymer concrete,
 corrosion in, 201
Microwave-oven curing, 93
Middle ages, 9–11
Mineralogical phases, 36
Ming Dynasty Great Wall, 9
Minitab software, 127
Modulus of resilience (MOR),
 PA-based GP mortar, 212, 215
Molding,
 PA-based geopolymer products,
  203, 204, 205
 PA/HCFC slag-based GP
  preparation, 157
 PA/jute fiber-based GP, 179, 180,
  181
 preparation of geopolymer from PA,
  99, 100
Monasticism, building techniques, 9
Mongolian pond ash, 93
Morphologies of geopolymer, 36, 94,
 148, 158, 171, 182, 206
 as-received pond ash, 141, 146
 coal ash, 33
 pond ash, 31, 32
 surface, of prepared materials, 101
Mortar(s),
 PA/jute fiber-based GP, 178–185
 preparation, PA-based geopolymer,
  203–204
Mud brick, 2, 4, 5

Mullite, 218
Museum of Prehistoric Thera in Santorini (Greece), 3

NALCO Navratna Company, 91, 98, 126, 156, 170, 175, 203
Nanchan Temple (Wutai), 8
Nardite, 49
Natural ennore sand, 93
Natural fibers, 170–172, 173, 175–176, 177
Natural pozzolan (PN), 49
Neolithic age, 2–3
Neolithic tools, 2
Nepheline, 218
New stone age, 2–3
Nineteenth century, constructional materials, 16–17
Nomograms, for mechanical properties of GP, 141–147
Non-destructive testing (NDT) technique, 100
Notre Dame Cathedral, Paris, 11
NTPC, 170

Oil, energy source, 21, 22
Oligomers, defined, 79
Olivine, 44
OPC. see Ordinary Portland Cement (OPC)
Open-circuit potential (OCP) method, for corrosion measurement, 200
Ordinary Portland Cement (OPC), 31, 46, 47, 53, 72, 228
 -based concrete, 154
 corrosion behavior of, 198
 environmental impact of, 228
 partial substitution of, 172
 production, 152, 196
 with SCBA, 202
 using HCFA, 202
Ores, defined, 39

PA. see Pond ash (PA)
Palatine Chapel, Aachen, 9
Pantheon in Rome, 6
Parthenon, 6
Passivity of embedded steel, 55
Paver blocks, geopolymer, 228, 229
Peaks (bands), 219–220
Peat wood, 171
Percussion cup method, 35
Perkin Elmer Pyris Diamond analyzer, 207
Permeability,
 of coal ash, 37
 coefficient of, 33
 intrinsic, 33
 porosity and, 33–34
Phosphogypsum (PG), 49, 74
Phreatomagmatic eruptions, 20
pH value, fly ash, 36
Pieter Bruegel the Elder, 14, 15
Pineapple fiber, 173
Plasticizer, water-soluble, 97, 98, 105, 106, 126, 157, 175, 181
Pollution,
 air, 29–30, 32, 32, 170, 231
 coal-based thermal power stations, 92
 control, 24, 29–30, 42, 47, 72, 228, 234
 environmental, 46, 72, 125, 152, 234
 groundwater, 32
 industrial, 234
 OPC and, 152
 problem, solving, 234
Polypropylene fibers, 173–174
Polysialates, 45
Poly-vinyl alcohol (PVA) fibers, 174
Pond ash (PA), 23, 29–31
 application areas of, 31
 -based GP products, corrosion of. see Corrosion, of pond ash (PA)-based geopolymer products

chemical and physical
  characteristics, 30
chemical composition of, 30, 95, 126
chemicals
  Sika (water-soluble plasticizer),
    97, 98
  sodium hydroxide, 97
  sodium silicate, 96
collection, 92, 156
development, HCFC slag-based GP.
  see High carbon ferrochrome
  (HCFC) slag-based GP
engineering properties of, 32
fly ash *vs.*, 30–31
geopolymer production from, 72, 73, 79
macrograph of, 95
management, importance of, 32–33
mechanical activation of, 93
PA/HCFC slag-based GP
  material, characterizations of,
    158, 159
  mortar and concrete, 158, 159
  preparation, 157
phases of, 126
physical properties of, 93, 96
physicochemical properties, 31
preparation of geopolymer from
  curing, 100
  molding, casting and compaction,
    99, 100
  raw materials, 98, 99
process of generating, 31
reactivity of, 92, 93
silica content of, 153
storage, 93
variable characteristics of, 32
Pond ash (PA)-based GP cementitious
  materials,
  compressive strength, 104
  experimental details, 94–114
  materials, 94–98
  overview, 92–94
  preparation of geopolymer, 98–100
  results and discussion, 104–114
  test methods, 100–104
Pond ash (PA)-based GPs, strength
  property of,
  experimental details, 126–127
  characterization of prepared
    samples, 127
  materials and method, 126
  preparation of geopolymer, 126
  overview, 124–126
  regression coefficients, calculation
    of, 128, 130–131, 133
  results and discussion, 127–147
  morphologies of as-received pond
    ash, 141, 146
  nomograms, 141–147
  significance coefficients, testing
    of, 133–147
Pond ash (PA)–jute fiber-based GP
  cementitious materials,
  experimental details, 175–189
  chemicals and materials, 175–178
  compressive strength tests,
    182–185, 186
  concrete preparation, 181
  curing process, 179, 180, 181–182
  molding, casting and compaction,
    179, 180, 181
  mortar and concrete, 178 185
  mortar preparation, 180
  preparation, 179
  raw material preparation, 178,
    179–180, 181
  results and discussion, 186–189
  test methods, 182–185
  microstructures of, 186
  overview, 170–175
Pond fly ash (PFA), 31–32
Porosity,
  PA-jute fiber-based GP, 174
  permeability and, 33–34

Portland cements (PC),
  alternatives to, 47–48, 73
  hydration characteristics of, 53
  OPC. *see* Ordinary Portland Cement (OPC)
  Pozzolana, 31
  production of, 19, 47, 73
  substitute material for, 38
  typical composition of fly ash, 28, 29
Portlandite, 49, 75
Potassium aluminate (KA), 156
Potassium metasilicate (KS), 156
Pourbaix diagram, 198, 199
Power-compensated DSC, 103
Pozzolanic effect of ashes, 29–30
Processed pond ash, 178, 179, 181
Processed sand, defined, 181
Protective coatings (PCs), geopolymer, 229–230
Protective equipment, use of, 17
Pulverized fuel ash, various uses of, 31–32
Pyrolysis process, 103

Quarry dust, 20
Quartz, 39, 114, 218

Rebar, rusting of, 54
Red mud (RM), 172
Regression coefficients, calculation of, 128, 130–131, 133
Reinforced concrete (RC) building materials, 201
Reinforced fibers, 173
Reinforced pond ash, shear strength of, 171
REMI mold, 99, 126, 179, 180, 181, 203–204, 205
Renaissance, the, 11–15
Residual waste glass (RWG), 49, 74
Rheims Cathedral, 9, 10
Rice husk ash (RHA), 19, 172, 173–174

Rock slag, 42
Romanesque style of architecture, 9
Rome, ancient, 5–8
Roofing, 12, 171
Rusting of rebar, 54

Salisbury Cathedral, 15
Sandy silts, 34
Santa Maria del Fiore, in Florence, 12, 13
Sawdust, 171
Scanning electron microscopy (SEM),
  HCFC slag-based GPs, 152, 153, 155, 156, 158, 159–161
  PA-based GP mortar/concrete, 206, 214, 217
  PA-based GPs, 101–102, 107, 108–110
  PA-jute fiber-based GP, 172, 182, 186
  zeolite-based geopolymer, 202
Self-cleaning concrete materials, GP-based, 231
SEM. *see* Scanning electron microscopy (SEM)
Sericite powder, 230
Seventeenth century, constructional materials, 15
Shear strength, of coal ash, 37
Shyam Ferro Alloys, 156
Sialate, 45
Significance coefficients, testing of, 133–147
Sika (water-soluble plasticizer), 97, 98, 105, 106, 126, 157, 175, 181
Silica, 19, 151, 153
  fume, 93, 174, 202
  polymorph, 79
Silica-alumina-bearing waste material. *see* Fly ash (FA)
Silico-aluminates, 45

Silicon-bearing materials, GP preparation, 154
Siloxane oligomers, 79
Silty sands, 34
Sindhunata, 155
Sisal and coir fibers, residual, 171
Skara Brae in Scotland, 2
Slag-based geopolymer concretes (SGPC), 154
Slag(s), 24, 39–45
  ACBFS, 42
  AOD, 48, 73
  BFS, 40, 41, 42, 44
  BOF, 42
  Ca electric arc ferronickel, 73, 76
  cement, 42, 44
  chemistry, 39
  converter, 44
  electric arc furnace slag, 44
  ferrous and non-ferrous smelting processes, 39
  GBFS, 42
  generation of, 40–44
  geopolymer production from, 72, 73, 74, 75, 76, 77, 78
  GGBFS, 42
  HCFC slag-based GP. *see* high carbon ferrochrome (HCFC) slag-based GP cementitious materials
  iron and steel slag, 43, 44
  LD or steel slag, 42, 44, 45
  metallurgical, 51, 77
  optical photo images, 41
  properties and utilization, 44–45
  reducing, 44
  rock, 42
  sodium sulfate-activated slag cements, 51, 77
  steel. *see* Steel slags
  synthetic, 39
Slave labor, use, 7
Smelting,
  copper, lead and bauxite, 39
  ilmenite, 39
  processes, ferrous and non-ferrous, 39
Soda-lime waste glass powders, 77
Sodium aluminosilicate hydrate, 218
Sodium carbonate, 114, 153
Sodium hydroxide (SH), 49, 50, 74, 76, 97
  from Loba Chemicals, 126
  preparation of geopolymer, 154, 156, 180, 181
  variations of, 104–105
Sodium perborate, 233
Sodium silicate (SS), 48, 49, 50, 73–74, 76
  chemical structure of, 96
  from Merck (India), 126, 156–157
  preparation of geopolymer, 154, 156, 180, 181
  variations of, 104–105
Sodium sulfate-activated slag cements, 51, 77
Sol-gel method, for geopolymers, 48, 73
Songyue Pagoda (China), 8
Specific surfaces, of coal ash, 35
Spray coating technique, 200
Sputter coater, 102
St. Paul's Cathedral, England, 15
Stave churches, in Scandinavia, 9
Steel,
  corrosion
    mechanism of, 54–55
    property of steel-reinforced concrete structures, 55
  dissolution mechanism of, 54
  mass production of, 16
  passivity of embedded, 55
  reinforcement, corrosion of, 198
  use, 16
Steel Age, 3–4

Steel slags, 42
  applications, 44
  compositions, 43, 44
  geopolymer production from, 72, 75
  powder, 50, 75
  production, 40
Strength properties of geopolymers, 48, 51, 53, 74, 77, 93, 104, 175
  at room temperature, 106
  compressive. *see* Compressive strength
  effect of parameters on, 115
  enhancement of, 153
  mechanical, 233
  mortars, 172
  of as-prepared composite, 171
  of PA-based geopolymer. *see* Pond ash (PA)-based GPs, strength property of
  proportion of FA/BFS blends, 153
  results and discussion on, 159–163
  volume fraction of glass fiber/polypropylene fiber and, 174
Subbituminous coal, 27
Sugar cane bagasse ash (SCBA), 202
Sulfuric acid, 93, 154
Super-sulfated cement (SSC), 49, 74
Supplementary cementitious materials (SCMs), 200
Surface water pollution, 32
Sweet sorghum, 171
Synthetic basalt, 94

Talcum powder, 230
Temple of Apollo at Didyma, 6
Temple of Vesta in Tivoli, Italy, 6
Tensile strength, of geopolymers, 22, 104, 176, 200
Ternary cement mortars (TBCMs), 172
Ternary ecological concrete (TEC), 202

Terracotta, 16
Tetra-calcium aluminoferrite, 44
Thermal conductivity,
  PA-jute fiber-based GP, 174
  TROLIT, 233
Thermal curing, 94
Thermal power plants,
  ashes, nature and composition of, 24, 26–29
  in India, 124
  pond ash produced in, 32
Thermal power stations,
  coal-based, 92
  pond ash and, 153
Thermal stability, 173, 175
  by TGA, 127
  of as-prepared PA-based GP mortar/concrete, 207
  of PA/Jute fiber-based GP, 182
  PA/HCFC slag-based GP, 158
  zeolite-based geopolymer, 201–202
Thermograms, DSC, 112, 141, 146, 147, 161, 162
Thermogravimetric analysis (TGA),
  HCFC slag-based GPs, 158, 159
  PA-based GP mortar/concrete, 207
  PA-based GPs, 102, 113, 127
  PA-jute fiber-based GP, 172
Thermo Nicolet NEXUS 870 FTIR spectrometer, 100
Titanium dioxide, 39
Tower of Babel, 14, 15
Trajan's column in Rome, 8
Treadwheel crane, Roman, 6, 7
Tricalcium silicate, 44
TROLIT, 233
Twentieth century, constructional materials, 17–18

Ultra-high-performance concrete, production, 155
Ultrasonic pulse velocity test, 201

UMT2 Micro Balance, 206
Undrained unconsolidated (UU) triaxial tests, 171
UNESCO World Heritage Site building, 12, 13
United States, coal deposits in, 21, 23
Uruk, 4

Vegetable fibers, 171
Villard de Honnecourt, 9, 10
Vitruvius, 11–12
Volcanic ash, 19–20

Waste(s),
  fibers in preparation of eco-friendly GP composite, 171–172
  glass-based GPs, 49, 74
  industrial
    bearing aluminosilicate minerals, 230
    dumping and landfilling, 170
    GPC from, 229
    GP production from, 72, 74, 76, 77
  materials, 229
    due to coal combustion, 92
  molding sands, 73, 76
  PA, dumping and landfilling, 170
  silica-alumina-bearing waste material. *see* Fly ash (FA)
Water impounded hopper (WIH) system, 29
Water-soluble plasticizer (Sika), 97, 98, 105, 106, 126, 157, 175, 181

Weight change, PA-based GP mortar/concrete, 207–211, 212, 213
Weight loss,
  of bricks, 174
  of GP cured samples, 146, 147
  measurement, 200–201
Wet method, 124
Wood ash, 19
Wooden building, Chinese, 8
Wood flour, 171
Wool, 171
Woolworth building, 16, 17
Wrought iron, use of, 16

X-ray diffraction (XRD), 94
  geopolymer cement using HCFA, 202
  HCFC slag-based GPs, 152, 153, 155, 156, 159
  PA-based GP mortar/concrete, 218
  PA-based GPs, 100, 101, 127
  PA-jute fiber-based GP, 172, 175
  zeolite-based geopolymer, 202
XRD. *see* X-ray diffraction (XRD)

Z300 Benchtop Centrifuge, 206
Zeolite-based geopolymer, 76, 80, 201–202, 218
Ziggurat of Ur, 4

Also of Interest

Development of Geopolymer from Pond Ash-Thermal Power Plant Waste

Novel Constructional Materials for Civil Engineers
Edited by Muktikanta Panigrahi, Ratan Indu Ganguly and Radha Raman Dash
Published 2023. ISBN 9781394166527

Utilization of waste materials has become a global challenge since they endanger our environment. In this book, the authors demonstrate how to utilize fly ash/pond ash (waste materials from thermal power plants) to produce a novel material called 'Geopolymer' (GP). Red mud, slags, etc., are mixed with fly ash to produce GP with enhanced strength. As shown in a few European countries, GP can replace cement, and some permanent structures constructed with GP are now appearing in a few advanced countries. GP, and geopolymer concrete, is considered suitable for the construction of roads, buildings, etc., and will eventually, fully or partially, replace cement.

The book highlights the mechanism of the formation of GP from pond ash. Properties of structures made with GP concrete are found to be comparable to those made with cement concrete. Systematic investigations are presented to understand the chemistry of GP formation with pond ash materials. Performances of these materials above ambient temperature, as well as with different environmental conditions, are also evaluated.

www.scrivenerpublishing.com

Printed and bound by CPI Group (UK) Ltd, Croydon, CR0 4YY
15/08/2023

03247073-0003